T0134427

Proceedings of the I-ESA Conferences

Volume 10

This series publishes the proceedings of the IESA conferences which began in 2005 as a result of cooperation between two major European research projects of the 6th Framework R&D Programme of the European Commission, the ATHENA IP (Advanced Technologies for Interoperability of Heterogeneous Enterprise Networks and their Applications, Integrated Project) and the INTEROP NoE, (Interoperability Research for Networked Enterprise Applications and Software, Network of Excellence). The I-ESA conferences have been recognized as a tool to lead and generate an extensive research and industrial impact in the field of interoperability for enterprise software and applications.

Bernard Archimède · Yves Ducq · Bob Young ·
Hedi Karray

Editors

Enterprise Interoperability IX

Interoperability in the Era of Artificial
Intelligence

 Springer

Editors
Bernard Archimède 🆔
École Nationale d'Ingénieurs de Tarbes
Tarbes, France

Yves Ducq 🆔
University of Bordeaux
Talence, France

Bob Young
Loughborough University
Loughborough, UK

Hedi Karray 🆔
École Nationale d'Ingénieurs de Tarbes
Tarbes, France

ISSN 2199-2533   ISSN 2199-2541 (electronic)
Proceedings of the I-ESA Conferences
ISBN 978-3-030-90883-6   ISBN 978-3-030-90387-9 (eBook)
https://doi.org/10.1007/978-3-030-90387-9

Organizers and Sponsors

Sponsors

Preface

Industry 4.0, smart cities, Internet of Things, big data, and digital transformation are the main paradigms and technologies of the artificial intelligence era. This era requires a foundation for seamless and secure communication called "interoperability". Moreover, the cooperation between different organizations such as manufactures, service providers, and government requires intelligent "Enterprise Interoperability" as well as applications and systems. Artificial intelligence (AI)-related technologies can determine the ability of systems to organize knowledge, make sense of it, increase decision-making and control, and extract value from data. Artificial intelligence has recently become a priority for many governments especially in Europe. However, it will only reach its full potential by overcoming the many barriers that remain to the interoperability of applications and computer systems.

Accordingly, Interoperability for Enterprise Systems and Applications (I-ESA2020) joins new intelligence models and trends technologies including IoT, cyber-physical systems, cloud computing, and digital twins. Connecting the world's leading researchers and practitioners of entreprise systems and applications, I-ESA2020 has presented outstanding exchanges of experiences and business ideas between researchers, services providers, entrepreneurs, and industrial stakeholders.

I-ESA2020 is the tenth of a series of conferences: Geneva (2005), Bordeaux (2006), Madeira (2007), Berlin (2008), Coventry (2010), Valencia (2012), Albi (2014), Guimaraes (2016), Berlin (2018) as well as a special edition in Beijing (2009), and this time is under the motto "Interoperability in the Era of Artificial Intelligence". The I-ESA2020 conference was hosted by Ecole Nationale d'Ingenieurs de Tarbes (ENIT) in France and jointly promoted by Pole Grand Sud-Ouest (PGSO I-VLAB) and European Virtual laboratory for Entreprise Interoperability (INTEROP-VLab—http://www.interop-vlab.eu). I-ESA2020 is technically sponsored by IFIP Workgroup on Enterprise Interoperability (5.8).

World leading researchers and practitioners in the area of entreprise interoperability contributed in this book. You will find contributions to interoperability solutions from different disciplines: computer sciences, engineering, and business administration.

The I-ESA2020 programme included several keynotes presented by high-level renowned experts from industry, government, and academia:

- Prof. Dimitryos Kiritsis, professor of ICT for sustainable manufacturing at EPFL, Switzerland
- Dr. Fernando Mas, Chief Technology Officer at COMLUX AVIATION, Indianapolis, USA
- Dr. Arian Zwegers, Programme Officer at European Commission, Brussels.

The book is organized into eight parts addressing major research topics in the scope of interoperability for entreprise systems and applications:

Part One: Ontology-Based Engineering
Part Two: Internet of Things
Part Three: Digital Twin
Part Four: Digital Platforms
Part Five: Processes Interoperability
Part Six: Model-Driven Approaches
Part Seven: Data and Knowledge Modeling
Part Eight: Business Oriented Applications

Tarbes, France   Bernard Archimède
Bordeaux, France   Yves Ducq
Loughborough, UK   Bob Young
Tarbes, France   Hedi Karray

Acknowledgments

We wish to thank all the authors, invited speakers, reviewers, senior programme committee members, and participants of the conference who made this book a reality and the I-ESA2020 conference a success; we express our gratitude to all organizations that have supported the I-ESA2020 preparation, especially INTEROP-VLab and the PGSO. We are deeply grateful to the local organization support notably Bernard Archimède, Hedi Karray, Raymond Houe-Ngouna, Linda Elmhadhbi, Maroua Masmoudi, and all the administrative staff of ENIT for their excellent work in the preparation and the management of the conference. We are proud and thankful to all the sponsors of I-ESA2020 especially the Region Occitanie, Agglomeration Tarbes Lourdes Pyrénées, EDF une rivière un territoire agency, and Comlux.

Contents

About the Editors

Prof. Bernard Archimède received the Ph.D. degree in Computer Sciences applied to industry from the University Bordeaux 1 in 1991. Since this date, he is researcher in the Laboratory Génie de Production (LGP) of the National School of Engineers at Tarbes (ENIT), where he is a full professor. His research deals with the distributed software architectures and their applications to the multisite scheduling and the control of complex systems. He was head of Dynamic Decision and Interoperability for Systems (DIDS) research team until September 2016. Since, he is deputy director of the Production Engineering Laboratory. He is also the president of the GSO pole (Grand Sud-Ouest) of the International Virtual Laboratory (INTEROP-VLab) for Enterprise Interoperability. Prof. Archimède has published more than 100 papers in international journals and conferences. He has broad interest on distributed planning architectures, distributed simulation, multiagent systems, and interoperability. He coordinated many research projects including a lot industrial collaborations.

Prof. Yves Ducq is a full professor and vice president of University of Bordeaux in charge of continuous improvement and documentation. Yves Ducq is doctor from University Bordeaux 1 in Production Management and Enterprise Modelling. He received his Ph.D. degree at the University Bordeaux 1 in 1999 and received his Accreditation to Supervise Research in 2007. He is working on Performance Measurement, Enterprise Modelling, Production Management, and Interoperability and has published more than 40 papers in books and international journals and more than 100 papers in international conferences. Prof. Ducq has been involved in several European projects for twenty years and particularly in the frame of IMS—GLOBEMAN 21 (FP4), Growth—EUROSHOE, IST—CENNET (cooperation with China), and IST—UEML of the FP5. He was strongly involved in INTEROP Network of Excellence and is now president of virtual laboratory on interoperability: INTEROP-VLab. He is also involved in many French research projects. He has also acted as research engineer on several contracts with industry on performance improvement and quality. In the frame of his vice presidency, he is in charge to spur projects related to continuous improvements, quality, process simplification, accreditations, and open science.

Prof. Bob Young is a visiting professor of Manufacturing Informatics in the Wolfson School of Mechanical, Electrical, and Manufacturing Engineering at Loughborough University in the UK. He is also the managing director of Redwood Informatics Limited, who provides informatics consultancy support to manufacturing industry. He has about 45 years' experience in new product development and manufacturing engineering, working both in the UK industry and in academia. Prof. Young's research has been funded directly by industry, by the UK funding agencies and by the EU. As well as support for industry, his work has led to around 200 research publications and some 20 Ph.D. completions. His research is focused on exploiting advanced Information and Communications Technologies to aid multi-disciplinary teams of engineers in their decision-making through the provision of timely, high-quality information and knowledge. To that end, his research in recent years has been heavily focused towards the development and use of formal ontologies as a basis for effective knowledge sharing and interoperability in manufacturing. Prof. Young works with a broad range of manufacturing companies from large multi-nationals in the aerospace and automotive sectors to more local manufacturing SMEs. In this latter area, he is a director of TANet, an organization aimed at providing support to the UK SME manufacturing sector. As well as working closely with industry, Prof. Young is committed to developing effective information standards for manufacture. To that end, he has been deputy convenor of the international standards organization working group concerned with "manufacturing process and management data", ISO TC184 SC4 JWG8. Prof. Young is the UK representative for the EU's Virtual Laboratory for Interoperability (INTEROP-VLab) and also leads a task group within this organization that is focused on developing improved methods for Manufacturing Enterprise Interoperability.

Prof. Hedi Karray is full professor of informatics at ENIT (National School of Engineers at Tarbes). He is the head of ontology research group at Production Engineering Laboratory. He received a M.Sc. degree in Information Technologies from University of Lyon in 2008, a Ph.D. degree in applied informatics, and an MBA from the University of Franche-Comté in 2012 then 2013 and the Habilitation to Lead Research (HDR) from the National Polytechnic Institute of Toulouse in 2019. In 2016, he had a visiting researcher position at the State University of New York at Buffalo (UB). Since his stay at UB, Dr. Karray became a senior scientist at the National Centre for Ontological Research. Hedi Karray is also a technical committee member of several international research groups such as IEEE SMC, IFAC 5.3, INTEROP-VLab and Industrial Ontologies Foundry. He has published more than 40 papers in international journals and conferences. He has served as a reviewer for several international journals and a member of several TPC in several international and national conferences. He chaired and organized the international conferences as IEEE AICCSA 2019 and I-ESA2020. Dr.

Karray has managed and participated in several collaborative research projects on the topics of Ontology-Based Engineering, semantic interoperability, and decision support systems. He has contributed to developing several ontologies and ontology-related tools in different domains such as engineering, crisis management, and astrophysics. In 2019, he was elected as IEEE senior member.

Ontology-Based Engineering

Toward Manufacturing Ontologies for Resources Management in the Aerospace Industry

Rebeca Arista, Fernando Mas, Carpoforo Vallellano, Domingo Morales-Palma, and Manuel Oliva

Abstract Manufacturing ontologies in the aerospace industry have been an active research topic during the last decade, as a mean for tool agnostic modeling and simulations activities supporting the product development process. This work reviews Models for Manufacturing (MfM) methodology, proposed by the authors to support ontologies generation in this field, as well as representative examples of its application. In addition, it proposes a preliminary ontology to support an activity of the development process in conceptual phase, in charge of generating a "Build Process" and "As-Planned" product structure. Special attention is made to the resource objects that take part of this process.

Keywords Aerospace manufacturing ontologies · Assembly systems ontology · Assembly line design · Knowledge-based systems · Models for Manufacturing (MfM)

R. Arista (✉)
Airbus, 2 Rond-Point Emile Dewoitine, 31700 Blagnac, France
e-mail: rebeca.arista@airbus.com

F. Mas
M&M Group, Av. Inventor Pedro Cawley 31, 11500 Cadiz, Spain
e-mail: fernando.mas@mecanizadosymontajes.com; fmas@us.es

F. Mas · C. Vallellano · D. Morales-Palma
University of Sevilla, Paseo de los Descubrimientos, s/n, 41092 Sevilla, Spain
e-mail: carpofor@us.es

D. Morales-Palma
e-mail: dmpalma@us.es

M. Oliva
Airbus, Avenida del Aeropuerto s/n, 41020 Sevilla, Spain
e-mail: manuel.oliva@airbus.com

© The Author(s), under exclusive license to Springer Nature Switzerland AG 2023
B. Archimède et al. (eds.), *Enterprise Interoperability IX*, Proceedings of the I-ESA Conferences 10, https://doi.org/10.1007/978-3-030-90387-9_1

1 Introduction

Models for Manufacturing methodology proposed by the authors [1] is based on a three-layer model (services, ontology, and data layers), enabling a horizontal integration against the traditional vertical system development of software vendors. It proposes a tool agnostic ontology layer to cover the knowledge of the company, linked to a data layer and supporting a services layer. This three-layer model approach would ease to migrate software applications inside the service layer with complete independence.

The data layer is the lower layer, which contains all the databases and interfaces: databases from the commercial software applications, legacy databases, clouds, data lakes, and many others, including the databases which hold the information instanced using the ontology layer.

The ontology layer is the middle layer, and it is the core of the model. It holds all the company processes, data, and semantic models, including the associated behavior and business rules. The ontology layer is the core in the three-layer model framework and is where the knowledge of the company is created, stored, managed, and used. The service layer is the upper layer, and it holds the software services: authoring, simulation, visualization, data analysis, among others.

The ontology layer is defined in an agnostic way; it is not linked to any model language or software tools. This layer includes the following models:

- **Scope model**. It defines the limits where the ontology works as the first step defining the ontology. It contains the main objects of the data model and a high-level definition of the system behavior.
- **Data model**. It defines the information managed in the selected scope, detailing all the data model objects.
- **Behavior model**. It defines the internal behavior of the systems on the given scope.
- **Semantic model**. Information coming from the databases is defined in different ways, different languages, or different formats (i.e., date in American or European format). This model defines the different semantics of the data model objects, to allow connecting to the data layer, instance the ontology with real data, and run real scenarios. The semantic model allows also to maintain connection between models among the lifecycle, providing digital continuity to the ontologies.

The MfM methodology promotes the use of a Model Lifecycle Management based on PLM systems, to support the collaborative process of building the ontology layer and managing the models and objects lifecycle, including configuration and effectivity. A PLM system can hold and easily manage ontology objects and let the engineers manage, upgrade, reuse, and enrich the objects. The authors have built an initial prototype based in a free and open-source software (FOSS) PLM [2].

2 Previous Work Done in Manufacturing Ontologies

This section shows representative research work made by the authors, applying Models for Manufacturing methodology and Model Lifecycle Management principles described in the previous section.

Prior to the explicit generation of this methodology and framework, the authors explored several knowledge-based engineering (KBE) implementations including an application using KBE to design and industrialize tools for high-speed milling machines (HSM) [3], demonstrating a methodology to represent the knowledge in a semi-structured way and the integration of general-purpose CAD/CAM systems.

A preliminary implementation of Models for Manufacturing to an innovative incremental sheet forming process was demonstrated in [4]. This manufacturing process makes use of computer numerical control (CNC) technology and usually requires a numerical study to validate the process. The work presented the ontology models under development, defined by means of IDEF0 diagrams and concept maps. Part of the conclusions of this work was that designing the system scope model among all project participants is an efficient method to establish the starting point for development of both ontology and software tools layers. Furthermore, the combination of more than one modeling techniques allows defining a robust ontology from different perspectives. A second application of MfM methodology was presented in [5], defining an ontology defining aerospace assembly lines in Airbus. The paper introduces a novel way to characterize the adherence concept for designing aerospace assembly lines. The MfM methodology, based on the three-layer model, allowed developing a scope model and enriching the data model with attributes and relations. FOSS modeling tools, Ramus and CMap, are used in this proof of concept, offering an easy way to establish models as well as for sharing and facilitating further discussion of the ontology. ARAS Innovator PLM system was used as Model Lifecycle Management (MLM) system to store the ontology and to manage lifecycle and configuration of different objects.

A preliminary model-based approach for gender analysis of Airbus Research Organization was the proof of concept developed in [6]. This work applied model-based systems engineering (MBSE) techniques to approach gender diversity, with the target to increase the women representation up to 30% in Airbus Research population from 2017 to 2021. The novelty of this work was creating an ontology as an analytical framework to approach the complex gender diversity social problem, proving the applicability as well of MfM methodology to other domains outside manufacturing. The ontology was composed by a scope model, a data model, and behaviors models, generated by use of several FOSS tools. An instance of the defined ontology was made using real historical data from Airbus Research Organization population, generating simulations to support discussions on potential action plans to take on this perimeter.

3 Preliminary Ontology for Collaborative Engineering Process Support

The iDMU concept (industrial Digital Mock-up), was defined by the authors in [7] as the unique deliverable of the collaborative engineering development process, including the product functional and industrial design, as well as the industrial system design. This concept sets the foundation for the proposed ontology framework presented in [8], to support the iDMU generation in the conceptual phase of development process.

This section describes a preliminary ontology as part of this ontology framework, supporting a development process type where an existing industrial system would be redesigned or reconfigured for an existing product. Inside this development process type, the scope is centered on the design process activity in charge of generating the so-called "Build Process" and "As-Planned" product structure [5].

The development process of a product and industrial system can be triggered in different combinations, being some of them: a new product and new industrial system design; a new product or derivate design with an existing industrial system redesign or reconfiguration; persisting product with an existing industrial system redesign or reconfiguration for performance improvement. The third type is addressed on this paper. In this work, the words manufacturing and assembly are used as per definition in the oxford dictionary; *manufacturing* is the making of articles on a large scale using machinery [9]; *assembly* is the action of fitting together the component parts of an object [10]. A "Build Process" describes for a given aircraft product, the high-level sequence of manufacturing, assembly, and logistics between plants, in a worldwide network. It considers the workshare to be made between countries, and it is the high-level definition of the process structure.

For a mono-configured product, the "As-Planned" structure is the industrial product definition, created from a layer of objects components of the "As-Designed" definition. The "As-Planned" defines the product work sharing between partners and subcontractors and their responsibilities. Every component in the upper level to the common layer defines a product responsibility of an assembly line, either for major components or for a final assembly line (FAL).

The "As-Planned" shows the product decomposition reflected in the assembly process stages of the "Build Process", from the elementary objects to the different subassemblies until reaching the complete aircraft. Multiple "As-Planned" structures fulfill different industrial scenarios maintaining an invariant "As-Designed" structure and therefore the same configuration layer of elementary objects.

The resource structure conforms the industrial system and comprises the complete footprint of resources and means used to produce the aircraft. These resources are huge industrial installations that involve complex assembly processes, sophisticated jigs and tools, machines and industrial means, and skilled human resources, including all the parts of the extended enterprise.

Different research works propose a resource structure and ontologies with points of view [11–13]. In this work, we consider only two higher levels of the resource structure decomposition: the global industrial system level and the plants and logistic means in a worldwide network level, which conform the global industrial system. The plant and logistic means at this level are treated as black boxes to the resource structure beneath them.

The next sections describe the scope model and data model following MfM methodology and an instance illustration of this ontology using the DA08 aircraft [5] and its global industrial system.

3.1 Scope Model

The scope of the proposed ontology is focused only on the activity "Build Process" and "As-Planned" definition of the development process type of industrial system reconfiguration, described by the IDEF0 model in Fig. 1. The elements considered from the product, process, and resource structure are only the ones in the lower level for the product and resource and the higher level of the process structure. These elements are called elementary objects, being a black box representation in the conceptual phase of the elements that will be designed in detail during definition phase.

The product elementary objects (PEOs) are black boxes of the product and joints between them (e.g., body, wing, joint body-wing, etc.), which create the common layer for the "As-Design" and "As-Planned" product structures at conceptual phase.

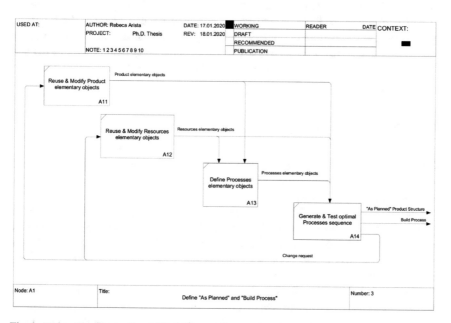

Fig. 1 Define "As-Planned" and "Build Process"

The resource elementary objects (REOs) are the resource black boxes at plant level with the attributes of capabilities and location (e.g., manufacturing plant with manufacturing capability at location 1, logistic mean with transport capability, etc.). Notions such as capability details, time, costs (recurring and non-recurring), among others, could be described for REO [14–16] but are not part of this work.

The process elementary objects (PrEOs) are the processes black boxes, which are defined in type by the REO capabilities (e.g., manufacture, transport, etc.) and in need by the PEO. The process precedence is considered only at this level of the network.

Activity A11 *"Reuse & Modify Product Elementary Objects"* and Activity A12 *"Reuse & Modify Resource Elementary Objects"* reuse the existing PEO of the "As-Designed" product structure and REO of the existing industrial system. The PEOs and REOs control Activity A13 *"Define Process Elementary Objects"*, where one PrEO is defined by each capability defined in the REOs. If one REO has a set of capabilities, one PrEO will be defined by each capability. The PEOs details will define all PrEO that will be needed to create the final product, regardless of the REOs capabilities.

The three set of objects (PEOs, REOs, and PrEOs) control Activity A14 *"Generate & Test Optimal Process Sequence"*. The PrEOs consume all PEOs and REOs in different possible scenarios. Different process sequences are generated with the scenarios of the PrEOs, considering rules and precedence constrains. One process sequence defines one complete "Build Process" scenario.

An optimal process sequence is selected against a defined criterion, being this the final "Build Process" scenario, with the PrEO sequence to be performed using the set of REOs, to the set of PEOs. This optimal sequence creates thus the "As-Planned" product structure with the PEOs attached to the PrEO sequence. If no optimal process sequence is selected, a "change request" can be generated to modify either PEO and REO or both, running again the activities until reaching a solution.

3.2   Behavior Model

The behavior model linked to the defined scope model is shown in the concept map in Fig. 2. Each statement is described by a triplet (subject, predicate, and object), using the sense of the arrows.

One PrEO has one REO and has at least one PEO. The PEOs define the needed PrEOs. A REO defines the type of PrEO by its capabilities; the nature of the capability will determine if one or multiple PEO can be allocated. The instances of PrEO objects will create a matrix or network type of link between PEOs and REOs instances depending on the type of PrEO.

The set of PrEO generates a set of process sequences considering rules and precedence constrains, which can trigger an optimization and analysis of those sequences. This can result on a change request on PEOs, with a different cut of the product and thus different set of PEOs. This can result as well on a change request on REOs,

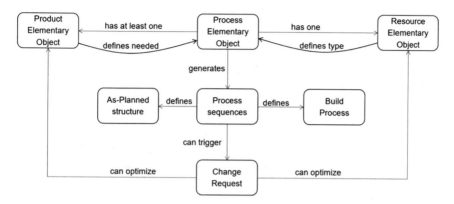

Fig. 2 Behavior model of "Build Process" and "As-Planned" definition

to consider a higher or lower number of resources, or different capabilities on each REO. An optimal process sequence selected defines the "Build Process" structure, with PrEOs sequence to be performed at the set of REOs to the set of PEOs, and defines the "As-Planned" product structure with the PEOs attached to the PrEOs sequence.

3.3 Instance on Use Case: DA08 and Global Industrial System

To illustrate the ontology models defined in this section, an instance is created with the DA08 artifact and a defined global industrial system for this artifact, used by authors as demonstration mean in several researches [5, 12].

First activities from the scope model are Activity A11 *"Reuse & Modify Product Elementary Objects"*, where manufacturing engineers reuse the product preliminary decomposition of the DA08 aircraft "As-Designed" structure in PEOs, in this case being: wing (PEO1), fuselage (PEO3), tail (PEO5), and the assembly drawings of the joint wing-fuselage (PEO2) and joint fuselage-tail (PEO4).

In Activity A12 *"Reuse & Modify Resource Elementary Objects"*, manufacturing engineers reuse the available or potentially available resources in quantity of a current industrial setup, including their capabilities and geographical location. The global industrial system defined in this case that could produce DA08 aircraft is formed by the REO components: one manufacturing plant with manufacturing capability at Kourou, French Guiana (REO1); two assembly plants with assembly capability one in Jacksonville, the USA, and another in Saint-Nazaire, France (REO2 and REO3); and one logistic mean with transport capability and non-applicable location (REO4).

In Activity A13 *"Define Process Elementary Objects"*, one PrEO is defined by each capability of the defined REOs. In this case, the REOs have only one capability which defines the following PrEO: manufacture by (PrEO1 of REO1), assemble by

(PREO2 of REO2, PrEO3 of REO3), and transport by (PrEO4 of REO4) illustrated in Fig. 3.

With the set of PEOs, the manufacture engineer defines the set of needs in terms of PrEO, to be able to deliver the complete product. A matrix is generated as shown in Table 1, with the PrEO instances linking PEO and REO for different scenarios. The index of the instances is defined by line/column, and in the case of an assembly process, the index contains an "x" and "y" for the products to be assembled and a "z" for the assembly drawings that defines the junction.

In Activity A14 *"Generate & Test Optimal Process Sequence"*, the optimization problem proposed is to find an industrial system configuration and process sequence that would produce the DA08 artifact in the shorter production time and with less non-recurring and recurring costs. This implies for the industrial system configuration to select from the two assembly plants available: if both plants are necessary

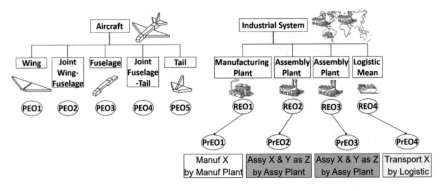

Fig. 3 Product, resource, process elementary objects of DA08 and global industrial system

Table 1 PrEO instances linking PEO and REO for scenarios

		PEO1	PEO2	PEO3	PEO4	PEO5
REO1	PrEO1_1	PrEO1_1.1	–	–	–	–
	PrEO1_2	–	–	PrEO1_2.3	–	–
	PrEO1_3	–	–	–	–	PrEO1_3.5
REO2	PrEO2_1	PrEO2_1.x.1	PrEO2_1.z.2	PrEO2_1.y.3	–	–
	PrEO2_2	–	–	PrEO2_2.x.3	PrEO2_2.z.4	PrEO2_2.y.5
REO3	PrEO3_1	–	–	PrEO3_1.x.3	PrEO3_1.z.4	PrEO3_1.y.5
	PrEO3_2	PrEO3_2.x.1	PrEO3_2.z.2	PrEO3_2.y.3	–	–
REO4	PrEO4_1	PrEO4_1.1	–	–	–	–
	PrEO4_2	–	–	PrEO4_2.3	–	–
	PrEO4_3	–	–	–	–	PrEO4_3.5
	PrEO4_4	–	PrEO4_4.2	–	–	–
	PrEO4_5	–	–	–	PrEO4_5.4	–

(affecting non-recurring cost); which of them should be the pre-assembly plant and which should be the final assembly plant; which PEOs should be assembled at each (affecting recurring cost). The process sequence selection would affect the total production time.

The three sets of objects (PEO, REO, and PrEO) control Activity *A14 "Generate & Test Optimal Process Sequence"*. PrEOs instances linked to REO should consume all PEOs. A first verification on the product split in the given PEOs is made at this stage, as the available resource capabilities will have to match the PEO product definition (e.g., material, technology, dimension, etc.). If this is not the case, a change request of the product split in the PEO can be requested, to have a new set of PEOs.

Once the PEOs match the REOs capabilities (at least once), different allocation options are done through PrEO instances for all possible scenarios. For the manufacturing plant (REO1), as it is the only resource available to manufacture, it will do so on the three product parts. In the same way, the logistic mean being the only one available (REO4) will have to transport all parts and joint parts if needed. Both plants can perform the two assemblies defined in the joints PEO2 and PEO4 and in different sequence.

All possible process sequence scenarios are generated with the instanced PrEO elements as shown in Fig. 4, considering rules and processes precedence constrains. These sequences are possible "Build Process".

An optimal process sequence is selected against the defined criterion (in this case, time and cost). This process sequence is the final "Build Process", with the PrEOs sequence to be performed using the set of REOs, to the set of PEOs, as shown in Fig. 5.

This optimal sequence creates thus the "As-Planned" product structure with the PEO in the common layer of the product structure attached to the PrEO sequence. Both "As-Design" and optimized "As-Planned" product structures are shown in Fig. 6. Additional product split in PEOs of the same artifact DA08 can be tested for different optimizations and/or REOs potentially available in the global industrial system with set of capabilities.

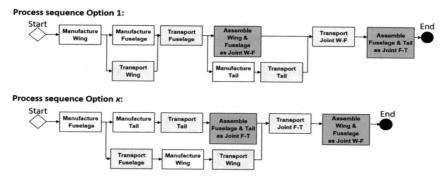

Fig. 4 Process sequences generated with different "Build Process" scenarios

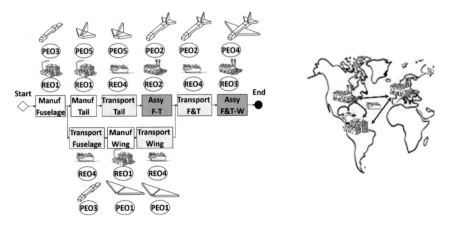

Fig. 5 Optimized "Build Process" with worldwide network

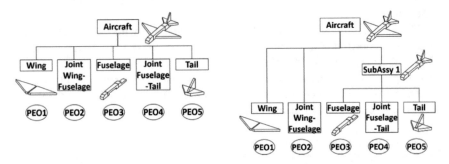

Fig. 6 "As-Designed" and optimized "As-Planned" product structures

4 Conclusions and Further Work

This work reviews Models for Manufacturing methodology proposed by the authors and different application cases of this methodology in the aerospace industry to define an ontology layer. A preliminary ontology is presented, to support a collaborative engineering development process to reconfigure an existing industrial system with a given product design, in the activity of "Build Process" and "As-Planned" definition.

A scope model and behavior model conforming this preliminary ontology are described, following an instance illustration for the use case of DA08 artifact and global industrial system. The illustration shows the benefits this would have for manufacturing engineers on this design activity at the conceptual phase of this type of development process. Further work will focus on the resource ontology considering capability details, time, costs (recurring and non-recurring), lifecycle, and granularity.

Acknowledgements The authors would like to thank Sevilla University colleagues, Airbus colleagues in France and Spain, and Comlux colleagues in the USA, for their support and contribution during the development of this work.

References

1. Mas, F., Oliva, M., Racero, J., & Morales-Palma, D. (2018). A preliminary methodological approach to Models for Manufacturing (MfM). In *Product lifecycle management to support industry 4.0. PLM 2018. IFIP advances in information and communication technology* (Vol. 540). https://doi.org/10.1007/978-3-030-01614-2_25.
2. Arista, R., Mas, F., Oliva, M., Racero, J., & Morales-Palma, D. (2019). Framework to support Models for Manufacturing (MfM) methodology. In *Proceedings of 9th IFAC conference on manufacturing modelling, management and control* (Vol. 52, pp. 1584–1589). https://doi.org/10.1016/j.ifacol.2019.11.426.
3. Rios, J., Jimenez, J. V., Pérez, J., Vizán, A., Menéndez, J. L., & Mas, F. (2005). KBE application for the design and manufacture of HSM fixtures. *Acta Polytechnica, 45*(3).
4. Morales-Palma, D., Mas, F., Racero, J., & Vallellano, C. (2018). A preliminary study of Models for Manufacturing (MfM) applied to Incremental Sheet Forming. In *IFIP international conference on product lifecycle management PLM18 proceedings* (pp. 284–293). https://doi.org/10.1007/978-3-030-01614-2_26.
5. Mas, F., Racero, J., Oliva, M., & Morales-Palma, D. (2018). Preliminary ontology definition for aerospace assembly lines in Airbus using Models for Manufacturing methodology. In *Proceedings of 7th international conference on changeable, agile, reconfigurable and virtual production CARV18*. https://doi.org/10.1007/978-3-030-01614-2_26.
6. Arista, R., & Mas, F. (2018). A preliminary model-based approach for Gender analysis of Airbus Research Organization. In *Proceeding of the IEEE international conference on engineering, technology and innovation*. https://doi.org/10.1109/ICE.2018.8436319.
7. Mas, F., Menendez, J. L., Oliva, M., Rios, J., Gomez, A., & Olmo, V. (2014). iDMU as the collaborative engineering engine: research experiences in Airbus. In *Proceedings of IEEE engineering, technology and innovation international conference*. https://doi.org/10.1109/ICE.2014.6871594.
8. Arista, R., Mas, F., Oliva, M., & Morales-Palma, D. (2019). Applied ontologies for assembly system design and management within the aerospace industry. In *Proceedings of the joint ontology workshops JOWO 2019, 10th international workshop on formal ontologies meet industry* (Vol. 2518). URN:nbn:de:0074-2518-1. http://ceur-ws.org/Vol-2518/paper-FOMI1.pdf.
9. LEXICO Oxford dictionary. *Manufacturing.* https://www.lexico.com/en/definition/manufacturing. Last accessed 2020/01/22.
10. LEXICO Oxford dictionary. *Assembly.* https://www.lexico.com/en/definition/assembly. Last accessed 2020/01/22.
11. Sanfilippo, E., Benavent, S., Borgo, S., Guarino, N., Troquard, N., Romero, F., Rosado, P., Solano, L., Belkadi, F., & Bernard, A. (2018). Modeling manufacturing resources: An ontological approach. In *Proceedings of the 15th IFIP WG 5.1 international conference PLM 2018* (Vol. 540). https://doi.org/10.1007/978-3-030-01614-2_28.
12. Mas, F., Rios, J., Menendez, J. L., & Gomez, A. (2013). A process-oriented approach to modeling the conceptual design of aircraft assembly lines. *The International Journal of Advanced Manufacturing Technology, 67*(1–4), 771–784. https://doi.org/10.1007/s00170-012-4521-5
13. Solano, L., Rosado, P., & Romero, F. (2014). Knowledge representation for product and processes development planning in collaborative environments. *International Journal of

Computer Integrated Manufacturing (pp. 37–41). https://doi.org/10.1080/0951192X.2013. 834480.

14. Chengying, L., Xiankui, W., & Yuchen, H. (2003). Research on manufacturing resource modeling based on the O-O method. *Journal of Materials Processing Technology, 139*, 40–43. https://doi.org/10.1016/S0924-0136(03)00179-1

15. Wiendahl, H. P., ElMaraghy, H. A., Nyhuis, P., Zäh, M. F., Wiendahl, H. H., Duffie, N., & Brieke, M. (2007). Changeable manufacturing-classification, design and operation. *CIRP Annals, 56*(2), 783–809. https://doi.org/10.1016/j.cirp.2007.10.003

16. Järvenpää, E., Siltala, N., Hylli, O., et al. (2019). The development of an ontology for describing the capabilities of manufacturing resources. *Journal of Intelligent Manufacturing, 30*, 959–978. https://doi.org/10.1007/s10845-018-1427-6

Transition from Work-As-Imagined to Work-As-Done Processes Through Semantics: An Application to Industrial Resilience Analysis

Francesco Costantino, Antonio De Nicola, Giulio Di Gravio, Andrea Falegnami, Riccardo Patriarca, Massimo Tronci, Giordano Vicoli, and Maria Luisa Villani

Abstract Increasing industrial resilience is a big challenge for manufacturing enterprises that are continuously facing severe accidents causing injuries, casualties, and economic losses. Assessing industrial resilience requires the analysis of production processes in order to find possible safety flaws. Sociotechnical process management suffers often from misalignments of process descriptions according to formal organization documents or manager views (Work-As-Imagined) and actual work practices as performed by sharp-end operators (Work-As-Done). Furthermore, existing modelling approaches leveraging on techniques such as process mining from digital traces cannot be used to solve such misalignments as these traces are often hardly available. In this context, we propose a computational creativity approach for a semantics-driven transition from Work-As-Imagined to Work-As-Done process models based on the functional resonance analysis method (FRAM). In particular, through formalized semantics, it will be possible to use automatic reasoning for

F. Costantino · G. Di Gravio · A. Falegnami · R. Patriarca · M. Tronci
Sapienza University of Rome, via Eudossiana 18, 00184 Rome, Italy
e-mail: francesco.costantino@uniroma1.it

G. Di Gravio
e-mail: giulio.digravio@uniroma1.it

A. Falegnami
e-mail: andrea.falegnami@uniroma1.it

R. Patriarca
e-mail: riccardo.patriarca@uniroma1.it

M. Tronci
e-mail: massimo.tronci@uniroma1.it

A. De Nicola (✉) · G. Vicoli · M. L. Villani
ENEA—Centro Ricerche Casaccia, via Anguillarese 301, 00123 Rome, Italy
e-mail: antonio.denicola@enea.it

G. Vicoli
e-mail: giordano.vicoli@enea.it

M. L. Villani
e-mail: marialuisa.villani@enea.it

B. Archimède et al. (eds.), *Enterprise Interoperability IX*, Proceedings of the I-ESA Conferences 10, https://doi.org/10.1007/978-3-030-90387-9_2

identification of criticalities and prioritization of normal work analyses. To this aim, we introduce some examples of rule patterns, inspired by typical data quality issues, which can be automatically applied to guide such a transition. An explorative case study on chemical cleaning for industrial process is presented to clarify the proposed approach.

Keywords Safety · Ontology · Rules · Functional resonance analysis method · Computational creativity

1   Introduction

According to Dinh et al. [1], resilience is the ability to recover quickly after an adverse event, and adequate safety management strategies can contribute to increase the resilience of industrial processes. In sociotechnical work systems, safety can be considered an emergent property due to the often non-predictable interactions among humans and technological components. Manufacturing enterprises aiming to increase industrial resilience have to deal with a variety of severe accidents causing injuries, casualties, and economic losses. Industrial resilience should be about anticipating such events or at least enhance response capacity.

One of the existing approaches to assess industrial resilience is to model and analyse production processes in order to find possible safety flaws, for instance, by means of simulation approaches [2]. For several manufacturing enterprises, processes are characterized by a high number of complex human activities and relations and by a low usage of process support technologies. Furthermore, process descriptions are usually derived from company documents (e.g. standard, procedures, notes) or from manager perspective (WAI: Work-As-Imagined). These views could be different from the actual work as performed, which are usually derived from exhausting interview sessions with sharp-end operators (WAD: Work-As-Done). Such misalignment hinders availability of reliable process models to be analysed. Furthermore, existing approaches as those in [3, 4] that reconstruct the WAD processes from digital traces through techniques such as process mining often cannot be used as these traces are either not available or they cover minimal parts of the overall process.

In such context, our aim is to define an approach to support the work of safety analysts in designing WAD processes starting from WAI descriptions. We propose a computational creativity approach for a semantics-driven transition from Work-As-Imagined to Work-As-Done process models. This approach does not intend to cut sharp-end operators out of this knowledge elicitation process. We rather provide a means to suggest possible process variations of the WAI models to safety analysts, through a facilitated interpretation of relevant WAD details. In particular, we propose to use the CREAtivity Machine [5], i.e. a software tool enacting automatic reasoning on a semantic representation of the WAI model and on an application ontology, to generate the above-mentioned possible variations. We introduce a list of examples of model transformation rule patterns, inspired by typical data quality issues, to be automatically applied to guide such a transition. We provide also a case study on chemical cleaning to clarify the main elements of the proposed approach.

The remainder of the paper is organized as follows. Section 2 presents the related work in the area. Section 3 briefly describes the functional resonance analysis method (FRAM) [6], a process representation method used as a core of the safety analysis. Section 4 describes a case study concerning manufacturing enterprises and related to chemical cleaning. Section 5 presents our approach for a semantics-driven transition from WAI to WAD models. Finally, Sect. 6 closes the paper with some considerations on this computational creativity approach to safety management and some future research directions.

2   Related Work

The work, as it is carried out in the situated reality of the sociotechnical systems (WAD), takes place according to patterns—that is, according to criteria of compromise efficiency accuracy [7]—with the aim of achieving a well-defined objective, in a particular context, producing consequences that can be unexpected and modify context and objective. The context of sociotechnical systems in general is such that [8]: the environment is different from the one imagined in the project; the objectives are multiple and changeable; needs are variable and unpredictable; resources have been degraded (e.g. staff; competence; equipment; procedures; time); and there is a system of constraints/penalties/incentives generally put in place. Any operator possesses an operating know-how of the work context, and WAD adaptations belong to such know-how, that usually becomes hardly detectable. In the event of an accident, operators can be usually blamed, when contrasting prescriptions. The same prescriptions are on the contrary ignored, if not even discouraged, if they can ensure productivity.

Its attainable version—the work as disclosed—is a partial representation, whose the analyst can make instrumental use, or simply it can be influenced by the presence of prying eyes (e.g. people may not feel comfortable, they may try to deceive the viewer), even unconsciously. Many distortions due to social pressure can distract the disclosed version of the work from its adherence to the work as a fact. Finally, a practitioner may know her/ his own work, but she/he does not know how much of it is being done by another practitioner. The WAI is similar to a unitary reductionist perspective; the WAD is made up of many complexity-oriented different views.

Both are partisan stories but, while in the WAI we are interested in knowing the interpretation given by the narrator, in the WAD, the ideal narrator should be objective and impartial. For such reason, the interview lends itself well to the WAI, while naturalistic observations in conjunction with complementary semi-structured interviews are a glimpse of the work to suit the WAD.

Naturalistic observations—besides being extremely expensive and time-consuming—do not protect the detection of the WAD from biases. In this sense, the use of IT applications can complement traditional investigation techniques. Due to the dependency of the WAD reliability on the observer/interviewer, developing an automatic or semi-automatic technique for data collection in collaboration with sharp-end operators may have the potential to generate relevant benefits in terms of WAD development.

Furthermore, most of the existing works related to business processes analysis [3, 4] face the issue of process models conformance checking, which can be considered similar to the problem of alignment between WAI models and WAD. However, they use process traces, which for social-technical systems are often hardly available. Hence, with respect to them, we propose an automatic support to suggest possible WAI process variations to safety analysts. At the best of our knowledge, such approach is unprecedented.

3   Basic Notions on the Functional Resonance Analysis Method

The functional resonance analysis method (FRAM) is a method of resilience engineering (i.e. the discipline that aims to engineer resilient sociotechnical systems) that, giving a functional description of the many activities involved, allows to effectively represent a work domain. The FRAM does not assume preemptively that there is a unique valid way to perform the work. Following the principles of resilience engineering [9], through its four principles (equivalence of failures and successes, approximate adjustments, emergence, functional resonance [6]), it acknowledges the variability of processes as an essential condition for adaptability, and therefore, for resilience as an emerging effect at the system level. The FRAM gives a functional description of the processes whose various agents perform many activities (i.e. functions in FRAM terminology). Such activities are usually tightly interrelated, implying interrelation among their variabilities as well. Each agent (both individual and collective) of the sociotechnical system usually regulates its own functions' variability in order to harmonize with other functions' variability. Sometimes the actions of individual agents—given their inevitable bounded rationality based on local (i.e. non-systemic) knowledge—may interact in an unintended manner, giving rise to emerging out-of-control variability phenomena, a condition also known as functional resonance.

The method itself is composed of four steps (excluding the so-called step 0, i.e. establishing the purpose of the analysis: risk assessment for proactive analysis or accident analysis for reactive analysis):

1.   To identify the functions of interest; i.e. to delimit the scope of the model, to establish which functions are in focus—and therefore which must be detailed in the foreground—and to establish which must remain on the background. In FRAM, a function can interact with other functions by links (i.e. so-called couplings) in a process (i.e. instantiation) establishing which functions are being performed, how they are connected and under which specific conditions. In a

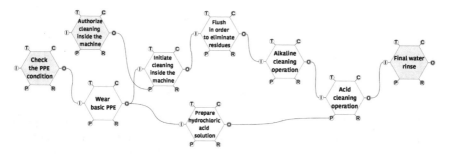

Fig. 1 FRAM model used for the analysis

single instantiation, the couplings link functions in sequential terms (i.e. an upstream function will precede a downstream function) and in modal terms; such mode is specified through the so-called six aspects; therefore, a FRAM function is traditionally depicted as a hexagon whose vertices are the aspects: Input (I), output (O), time (T), control (C), precondition (P), resource (R), see Fig. 1.

2. To identify the functions' variability. The variability of an activity is partly endogenous (intrinsic to the nature of the function itself), partly exogenous (specific to the context in which it is carried out), partly due to the specific upstream–downstream relationship that has taken place in the instantiation process. The entirety of these three components manifests itself at the output of each single function through the so-called phenotypes (i.e. the observable manifestations of variability at function level). The result of this step is the characterization of the potential variability as performed in the work context.

3. To aggregate the variabilities and, thus, to determine the actual variability. This step focuses on how the system affects, and it is in turn affected by, all the variability couplings, by the whole upstream–downstream interaction. Such intertwined functional aggregate determines the instance. Changing the scenario will produce another instance. By changing functions (in number, connected aspect, potential variability), another instance is obtained. Each possible variant begets a different FRAM instantiation. These instantiations can be used either for risk or accident analysis purposes. Moreover, FRAM allows comparing Work-As-Done and Work-As-Imagined simply by analysing the corresponding FRAM instantiations.

4. To manage variability. Since variability is necessary for the system to operate, it must be managed, not necessarily just damped, according to the scenario, by adequate work practices and possibly through suitable indicators.

4   Case Study: Chemical Cleaning Process

In this Section, we present a WAI description of a fragment of a chemical cleaning process, which is intended as a use case relevant for safety analyses of a typical manufacturing process. The work domain is described by means of the FRAM notation presented in Sect. 3.

The fragment depicted in Fig. 1 represents the following scenario. A sharp-end operator initiates cleaning operation inside the machine after he/she receives the authorization by the production manager. Such authorization is a precondition to start the cleaning operations. Meanwhile, the operator checks its personal protective equipment (PPE) and, wears it, if not ready yet. The output of the activity *wear basic PPE* is an input of the activities *initiate cleaning inside the machine* and *prepare hydrochloric acid solution*. Once the cleaning activity is started, the sharp-end operator flushes the machine in order to eliminate residues. Then he/she performs the alkaline cleaning operation and afterwards the acid cleaning operation. This operation needs as a precondition that a hydrochloric acid solution is prepared beforehand. Finally, the sharp-end operator rinses the machine with water.

5   Semantics-Driven Transition from WAI to WAD

We present an approach where semantics-based techniques are used to drive the modelling activity of a FRAM analyst in the transition from a WAI model to a WAD model. In particular, we define a method aiming at supporting the analyst in the exploration of potential modelling alternatives of the WAI to identify those variants that may lead to WAD models. The variants that seem most promising to the safety analyst, based on his/her experience, are then evaluated by eliciting specific information from the sharp-end operators.

This approach follows ideas and goals of computational creativity, a subfield of artificial intelligence aiming at defining computational systems that create artefacts and ideas [10]. Generally, computational creativity methods address the problem of thinking something new, e.g. a risk situation, by varying one or more aspects of what already exists, e.g. old experiences of incidents or normal situations.

The proposed approach essentially consists of a human-in-the-loop generative method of FRAM models, guided by automatic reasoning techniques that leverage on: the semantics of the model components expressed by an ontology structured according to the FRAM conceptual elements, and a set of predefined logical rule patterns, representing recurrent misalignments between WAI and WAD.

5.1 Evolution of FRAM-Based Manufacturing Ontology

The FRAM-based manufacturing ontology gathers both application and domain knowledge structured according to the FRAM Upper-level Model (FUM), an upper model derived from the FRAM method that was initially discussed in [11]. Such knowledge concerns WAI processes, existing standards and domain ontologies on manufacturing and expert knowledge. The FRAM upper-level model gathers the most relevant FRAM concepts and the ontological relationships linking them. FRAM_Element is the generic concept of the FUM upper-level concepts that is specialized in agent, aspect, function, and phenotype. Then coupling allows representing how two different functions link together and Coupling_effect models the corresponding effect, which could be amplifying, damping and No_effect. The FUM relationships are modelled in the ontology as object properties. The hasAspect object property relates two Aspects. It is specialized in the hasControl, hasInput, hasOutput, hasPrecondition, hasResource, and hasTime object properties. hasFunction is the inverse relationship of hasAspect. The hasPhenotype object property relates an output with its phenotype. The hasDownstreamAspect object property between coupling and input and hasUpstreamAspect object property between coupling and output allow to specify the role of the aspects in a coupling. Finally, the hasEffect object property relates the coupling concept with the corresponding CouplingEffect.

The FRAM-based manufacturing ontology is built by means of an incremental approach starting from an automatic export from the FRAM WAI process to the FUM application ontology, which organizes the FRAM WAI elements according to the FUM upper model ontological entities. Then, the FUM application ontology is enriched by considering existing standards and domain ontologies and by involving experts [12]. This final step is fundamental both to transform the tacit implicit knowledge of stakeholders and sharp-end operators in new concepts and relationships and to further validate existing knowledge. A sketchy representation of this process is depicted in Fig. 2, which conceptualizes the FRAM-based manufacturing ontology building process, through the FRAM WAI, standards, existing manufacturing domain

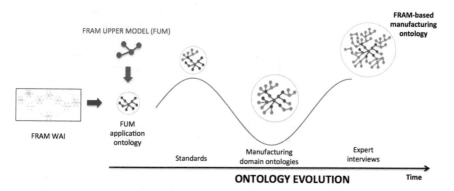

Fig. 2 FRAM-based manufacturing ontology evolution

ontologies as those addressed by the Industrial Ontologies Foundry group [13] and expert interviews.

5.2 Rule Patterns for WAI to WAD Transition

Given a WAI model, whose semantics is obtained by means of the FRAM-based manufacturing ontology, possible variations of model elements are generated by applying logical rules in queries to the ontology. These rules instantiate predefined model transformation rule patterns founded on some data quality dimensions proposed in [14]. The problem of misalignment of WAI models with the WAD may be faced as a problem of information quality occurring in the WAI models in their aim to effectively describing the WAD. With this meaning, the model transformation rules attempt information quality improvements of the WAI models.

We defined a patterns-based classification of model transformation rules, where a rule pattern may be founded on one or more data quality dimensions. As a preliminary outcome, we selected some data quality dimensions from the classification proposed by Pipino et al. in [14] and analysed them for the case study. In Table 1, we report the chosen quality dimensions and present some related transformation rule patterns. Each pattern is described by its purpose in the verification of the corresponding quality dimension over the model and by example rule types. One or more rules will be instantiated at run time with specific components of the model and enacted by means of queries to the FRAM-based ontology.

Table 1 Selected transformation rule patterns

Quality dimension	Description	Selection of transformation rule patterns
Completeness	The extent to which information is not missing and is of sufficient breadth and depth for the task at hand	*Purpose*: Verify conceptual representation coverage of model elements *Rule type*: **If** *model component* **is** a leaf concept **and** has a sibling, **replace** it with one of its siblings
Understandability	The extent to which information is beneficial and provides advantages from its use	*Purpose*: Verify appropriateness of model elements *Rule type*: **If** *model component* **is not** a leaf concept, **replace** it with one of its leaves
Relevancy	The extent to which information is applicable and helpful for the task at hand	*Purpose*: Verify organizational constraints of model elements *Rule type*: **If** *precondition* of function **is of type** *general organization rule,* **then remove** it from the model

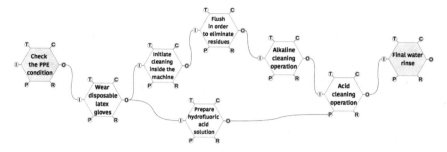

Fig. 3 FRAM WAD process

We show how the presented transformation rule patterns may be applied to the WAI model in Fig. 1 to suggest the model in Fig. 3, which better represents the WAD after evaluation of the information by the sharp-end operators (note a coupling is missing with respect to *initiate cleaning inside the machine*).

Transformation rule patterns related to *completeness* aim at verifying whether the WAI model contains all safety-relevant details of the real process. One method consists in checking whether a safety-relevant function, such as *prepare hydrocloric acid solution* in Fig. 1, correctly represents the practice. Indeed, the acid resource specified in the function description could not be available and sharp-end operators could use a similar type of acid instead. The attached rule type in Table 1, instantiated with the concept *prepare hydrochloric acid solution* as model component, would propose alternatives for that function, automatically retrieved from the ontology by means of concept similarity metrics. Thus, sharp-end operators could indicate the concept solution *prepare hydrofluoric acid* as the correct substitution for WAD model representation. It is worthy to note that, as such type of acid requires to be handled with special care, this could lead to safety flaws that deserve to be analysed.

Transformation rule patterns related to *understandability* aim at verifying whether the WAI model is correctly understood. One method consists in checking whether a safety-relevant function, such as *wear basic PPE* in Fig. 1, is not too generally described. Indeed, this level of abstraction of a function description would mean that sharp-end operators could choose any type of PPE, whereas some types of acid solutions may require specific PPE. The attached rule type in Table 1, instantiated with *wear basic PPE* as model component, would propose all the most specific variants for that function, retrieved from the ontology using concept subsumption relations. Thus, sharp-end operators would indicate the leaf concept solution *wear disposable latex gloves* as the correct substitution for WAD representation. However, as handling acid solutions could require wearing heavy chemical resistant gloves, this case deserves to be analysed in details.

Transformation rule patterns related to *relevancy* aim at verifying whether the WAI model is not over specified compared to WAD. One method consists in checking whether some preconditions are really required and do not block necessary functions, such as *initiate cleaning inside the machine* in Fig. 1, which requires *authorize cleaning inside the machine* to be performed first. However, in real work practices, some organizational procedures could be simplified, for example, to handle unexpected situations. Thus, sharp-end operators would confirm whether the concept solution *wear disposable latex gloves* is relevant for the WAD representation. Again, removing such function could lead to emergent issues that require to be explored by more detailed work domain analyses.

6 Conclusion

Industrial resilience of manufacturing enterprises requires anticipating accidents or improving the response capacity. A precondition to this is achieving a better understanding of industrial processes by means of an in-depth analysis. However, this is usually hindered by misalignments between WAI descriptions and WAD processes. In this context, we propose an automatic approach based on computational creativity and enhanced by semantics to support transition from WAI models to WAD models. To this purpose, possible WAI model variations are generated by applying some transformation rules according to patterns inspired by typical data quality issues and suggested to safety analysts. An exploratory case study on chemical cleaning shows an application of the method. To the best of our knowledge, application of computational creativity techniques to solve this misalignment problem is unprecedented. Future work will be devoted to further improve these techniques and to extend the list of transformation rule patterns.

Acknowledgements This work is partially supported by the $H(CS)^2I$ project, which is funded by the Italian National Institute for Insurance against Accidents at Work (INAIL) and Institution of Occupational Safety and Health (IOSH) under the 2018 SAF∈RA EU funding scheme.

References

1. Dinh, L. T. T., Pasman, H., Gao, X., & Mannan, M. S. (2012). Resilience engineering of industrial processes: Principles and contributing factors. *Journal of Loss Prevention in the Process Industries, 25*(2), 233–241.
2. Mourtzis, D., Doukas, M., & Bernidaki, D. (2014). Simulation in manufacturing: Review and challenges. *Procedia CIRP, 25*, 213–229.
3. Bloemen, V., van Zelst, S., van der Aalst, W., van Dongen, B., & van de Pol, J. (2019). Aligning observed and modelled behaviour by maximizing synchronous moves and using milestones. *Information Systems*, pre-print.
4. Lee, W. L. J., Verbeek, H. M. W., Munoz-Gama, J., van der Aalst, W. M. P., & Sepúlveda, M. (2018). Recomposing conformance: Closing the circle on decomposed alignment-based conformance checking in process mining. *Information Sciences, 466*, 55–91.
5. De Nicola, A., Melchiori, M., & Villani, M. L. (2019). Creative design of emergency management scenarios driven by semantics: An application to smart cities. *Information Systems, 81*, 21–48.
6. Hollnagel, E. (2012). FRAM: The functional resonance analysis method—modelling complex socio-technical systems, *Ashgate.*

7. Hollnagel, E. (2010). The ETTO principle: efficiency-thoroughness trade-off—Why things that go right sometimes go wrong. *Risk Analysis, 30*, 153–154.
8. Shorrock, S. (2016). The varieties of human work | humanistic systems, Steven Shorrock blog—Humanist System, 1–10.
9. Woods, D. D., & Hollnagel E. (2006). Prologue: Resilience engineering concepts. In E. Hollnagel, D. D. Woods, & N. Leveson (Eds.), *Resilience Engineering: Concepts and Precepts* (pp. 1–6). Ashgate Publishing.
10. Colton, S., & Wiggins, G. A. (2012). Computational creativity: The final frontier? In *Proceeding of the 20th European Conference on Artificial Intelligence*, pp. 21–26.
11. De Nicola, A., Vicoli, G., Villani, M. L., Patriarca, R. & Falegnami, A.: Enhancement of Safety Imagination in Socio-Technical Systems with Gamification and Computational Creativity. In *Enhancing Safety: The Challenge of Foresight. Proceedings of the 53rd ESReDA Seminar Hosted by the European Commission Joint Research Centre* (pp. 158–169) 14–15 November 2017, Ispra, Italy.
12. De Nicola, A., Missikoff, M., & Navigli, R. (2009). A software engineering approach to ontology building. *Information Systems, 34*(2), 258–275.
13. Industrial Ontologies Foundry (IOF) website, https://www.industrialontologies.org. Last accessed 7 November 2019.
14. Pipino, L. L., Lee, Y. W., & Wang, R. Y. (2002). Data quality assessment. *Communications of the ACM, 45*(4), 211–218.

Knowledge Extraction for the Product Development Process Based on Ontology-Driven Semantic Interoperability

Athon F. C. S. de Moura Leite⑩, Matheus B. Canciglieri⑩, Anderson L. Szejka⑩, Osiris Canciglieri Junior⑩, and Robert I. M. Young

Abstract The current product development scenario challenges manufacturing industry to deliver improved products to the market while ensuring improved quality. To ensure the best value, companies need to share product requirements effectively from various sources and domains, but there are still misinterpretation and mistakes on this process, regarding semantic interoperability obstacles in the context of the process of requirements' gathering, translating, and reusing them. To help in solving these problems, this study proposes an approach to aid the gathering of product knowledge, extracting it, and translating the knowledge for further use along the Integrated Product Development Process (IPDP). The approach consisted of analysing current issues of the topics, followed by the development of a novel approach, to then be further tested in an experimental case. Issues found on literature point to research gaps related to semantic reconciliation and the extraction of knowledge perspectives, in which semantic issues are approached through different points of view and multiple domains. The proposed approach considers unprocessed product requirements, further translated in features, and by that refining product knowledge during IPDP and enabling it to be reusable. The proposed solution shows a new method to collect and translate product requirements, while gathering its knowledge and transforming it in product features. The tests in an experimental case have shown a

A. F. C. S. de Moura Leite · M. B. Canciglieri · A. L. Szejka (✉) · O. Canciglieri Junior
Industrial and Systems Engineering Graduate Program (PPGEPS), Pontifical Catholic University of Paraná (PUCPR), Curitiba, Paraná, Brazil
e-mail: anderson.szejka@pucpr.br

A. F. C. S. de Moura Leite
e-mail: athon.leite@pucpr.edu.br

M. B. Canciglieri
e-mail: matheus.canciglieri@pucpr.edu.br

O. Canciglieri Junior
e-mail: osiris.canciglieri@pucpr.br

R. I. M. Young
School of Mechanical, Electrical and Manufacturing Engineering, Loughborough University, Loughborough, Leicestershire, UK
e-mail: r.i.young@lboro.ac.uk

B. Archimède et al. (eds.), *Enterprise Interoperability IX*, Proceedings of the I-ESA Conferences 10, https://doi.org/10.1007/978-3-030-90387-9_3

27

reduction in development time, and an increase in product quality, having significant impacts in reducing costs of development while ensuring correct communication and effectively sharing information.

Keywords Semantic web approaches · Knowledge management · Reference ontologies · Standardisation management and strategies

1 Introduction

Current product development is marked by complex requirements, higher standards of quality, and products that constantly need to adapt to fulfil customer's needs. This dynamism is tied to the trends on integrated manufacturing systems and Industry 4.0. The latter accompanies the use of technologies and methods to improve information sharing considering multiple domains of knowledge [1]. In this context, an Integrated Product Development Process (IPDP) deals with multiple domains of knowledge as a way to gather requirements to product development. The product requirements need to be well defined and shared with little or no loss of meaning during IPDP, in order to avoid misinterpretation, incoherency, and other issues during development, as increased costs and delays [2].

The contemporary practise is still not coping with those issues, that are semantic in nature, within a multiple domain environments, as product requirements must be consistent, clear, stand alone, measurable, testable, unique, unambiguous, and verifiable [1]. A significant portion of the issues in the requirements are related to them having different taxonomies for their representation, different points of view from agents during development, and limitations regarding the process of translating the knowledge in product development requirements. As a result, the misinterpretation of product requirements is related directly to wrong assumptions based on different information from heterogeneous domains [3].

In order to address the aforementioned issues, semantic interoperability (SI) has as its objective the effective information sharing in collaborative environments based on heterogenous domains. SI is being applied in the domains product design and manufacturing, in order to reduce semantic issues and cope with different sources of information [3]. Problems still persist, though, regarding their implementation, more specifically within the methods to extract and translate product requirements from heterogeneous sources of information, as well as standardising them [3].

This research has as its main objective the development of an approach to solve those issues by gathering, organising, and translating standardised product requirements and by that reducing the heterogeneity of interpretation during the IPDPD within a heterogeneous domain environment.

2 Conceptual Background

2.1 Integrated Product Development Process

Current IPDP relies on heterogeneous domains of knowledge, involving agents from different backgrounds and varied experience. Authors [4–7] cite ideal development cycles as integrated, collaborative, and interoperable, since as long as the product information is well defined, the misinterpretation of information and semantic barriers occurrence will be reduced [7].

Current models emphasise the systematisation process of IPDP, which was originally depicted as linear, with subsequent activities starting right after the conclusion of their predecessor [3]. Current approaches redefine that linear structure by inserting the notion of parallel activities across the product development [5, 7].

Research points to the necessity of ensuring interoperability in product and manufacturing in IPDP, as misinterpretation issues happen while product development requires multiple knowledge domains [4]. Research found in [8–10] presented the potential to use ontological methods to formalise knowledge in product and/or manufacturing models.

2.2 Ontology-Driven Semantic Interoperability

The use of ontologies has increased the development of shared representations. Recent research, as depicted in [9, 10], shows that the ability for sharing semantics across product and manufacturing representations can be supported by ontological formalisms. Ontologies are recognised as an important technology to cope with semantic interoperation issues [11]. Its formal structure provides machine-processed semantics of varied knowledge sources [12].

Despite their contributions, even when ontology-based methods are used, in order to assure shared semantics, semantic heterogeneity and their related issues are still unavoidable. Because of that, methods for proper ontology mapping are being developed to improve the semantics between ontologies representing domains that need interoperation [13].

Ontologies may be categorised in three distinct levels of abstraction during their application, depending on their aim [9]:

- Foundation Ontology, which is an ontology that is suited for general concepts and relationships, usable in heterogeneous domains;
- Reference Ontology, which is domain-specific ontology, being reusable in the same domain to perform different tasks; and
- Application Level Ontology, which represents knowledge that is specific and dedicated to unique tasks.

Fig. 1 Methodological procedures

3 Materials and Methods

This research uses as methodological procedures a qualitative literature review and an experimental case. Firstly, a literature review addressing the main issue and its dimensions is proposed, to identify the knowledge gap in which the solution will be developed. Later phases regard the creation and explanation of the approach steps, as depicted in Fig. 1.

The approach will be tested through an experimental case in which a real product will be scrutinised and compared considering its real development metrics and the ones obtained with the implementation of the proposed approach.

4 Literature Review

4.1 Cross-Domain Issues

In past decades, a few models for representation of standardised information structure were developed in heterogeneous domains. For instance, Unified Modelling Language (UML), Domain-Specific Language (DSL), and others [14, 15]. While presenting a way to formalise knowledge representation in different domains in a standardised manner, these models are not able to cope with the dynamic nature of product's requirements and knowledge from different phases of IPDP in a semantically accurate way [16].

Recently, a few models are considering the consistency of requirements and performance in environments with dynamic requirements. Authors in [17] explain a framework to support the design of cyber-physical systems, using design rationale and linking various system parameters and requirements coming from different sources. As shown in [18], authors investigate the manufacturing domain and the process of requirement gathering in different domains. The research in [17] and [18] combined different models to obtain verifiable and valid information within the context of dynamic requirements. However, the final word from specialists is still used remarkably, while translating these requirements in both models. This praxis might result in issues that are semantic in nature, as there might be significant subjectivity in the methods in which each specialist decides, due to their different comprehension of a domain.

4.2 Cross-IPDP Phase Issues

Communication in IPDP is based on the semantic interpretations of each agent [19]. In different phases of IPDP, heterogeneous sets of information might cause misinterpretation due to different meanings for a single term. That is a result of the different background of agents (e.g. product design, engineering, etc.) and their different levels of experience in the stages of product development [16, 17]. Authors in [16] state that knowledge that is required for a single product development stage might have different impacts in later activities, due to the dynamic nature of IPDP and the heterogeneity among their agents.

Currently, as shown in [18], research proposes a formalisation through semantic annotations for applications to semantically interoperate. However, there are no annotations that represent dynamic requirements, as well as automated ways of extracting them. In [20], the author presents an approach to solve cross-IPDP issues based on an ontology that is model driven, but exclusively to crossing two domains.

4.3 Cross-Requirement Issues

Requirements represent the main input in an ontology-driven semantically interoperable system related to IPDP. Their representation needs to be "semantically whole", in order to avoid negative issues, by using clear and well-defined axioms and statements [21]. Despite that, the poor abstraction of statements, in most cases, ends up generating interpretations that are divergent. This results in negative effects related to comprehension, uniqueness, and, in a significant portion, traceability of information.

In semantic interoperation, the comprehensible, unique, and traceable information is able to prevent inconsistencies in the product development and its manufacturing. In [21], a framework to cope with the issues and enable semantic interoperability is presented, in accordance with the previous statement. However, this framework does not ensure the traceability of the requirements and no optimisation of the process and structuration method of knowledge gathering. In [22], the authors presented a model that considers multiple domains, ensuring the traceability of information through verification and validation methods, but limited only to early phases of the development of a product. Current research, as shown in [1, 18, 21], shows the necessity to standardised procedures to extract information and knowledge, ensuring traceability through validation and verification, however, not considering the extraction of requirements. In [1] and [23], authors point out that future interoperable representations must consider knowledge extraction to ensure standardised knowledge gathering. In this sense, the proposed knowledge gap relates to "an automated product knowledge extraction in a multi-domain and interoperable environment that standardises knowledge in a holistic approach to IPDP, avoiding semantic issues".

5   Approach

In order to aid in the process of filling the exposed knowledge gap, an approach to increase automation in the process of gathering, extracting, and translating product knowledge into reference ontologies is proposed in Fig. 2, using IDEF0 notation. Such approach has as its goal the application of concepts and tools of semantic interoperability, in order to develop an interoperable environment. The concept is that this environment is able to represent and further translate knowledge among different phases of IPDP, analysing its consistency and reducing the negative effects caused by heterogeneous knowledge sources.

The approach is an extension of the Interoperable Product Design and Manufacturing System (IPDMS) model proposed in [1], considering its limitation of the process of knowledge gathering and extraction, and being used to develop its Reference View. The first phase, "Knowledge Gathering" (APKE1), consists in gathering knowledge from various sources in IPDP, i.e. Customer Relationship Management (CRM) information, Quality Function Deployment (QFD) information, Computer-Aided Design (CAD) drawings, Computer-Aided Manufacturing (CAM) information, and Computer-Aided Engineering (CAE) simulations.

Sequentially, the "Knowledge Pooling" (APKE2) occurs through a software application that reads the gathered information and extracts its features into an ".xml" extension file that represents the hierarchy of information and their properties. This software is a specialist software, referenced in this paper as "Approach for Product Knowledge Extraction System" (APKE-Sys) that must be developed considering its specific context, the organisation in which the approach is applied.

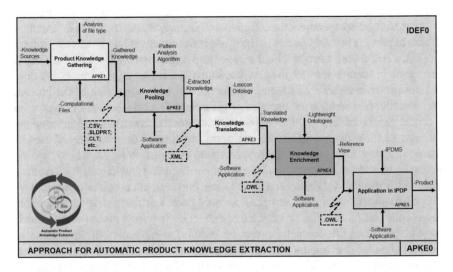

Fig. 2 Proposed approach

Before extracted knowledge can be used "safely", it must be translated and adapted in a way that ensures semantic integrity. This proves that there is still a problem regarding the terminologies used in companies, which may not be in agreement with the literature and/or research. To the translation to occur, a lexicon of the terms from the enterprise is used in the "Knowledge Translation" (APKE3) phase. The lexicon owns enterprise's concepts of products, the context associated with these concepts and their explicit meaning, being formalised in an ontology. The information gathered from the product is translated into information that is useful to aid product development, where the gathered information in ".xml" extension is converted to an ontology format (.owl) and compared to the lexicon ontology through ontology mapping algorithms in the APKE-Sys. The result is an ontology that contains the translated knowledge. The translation process occurs through mapping and intersection using a three-level similarity analysis done by the APKE-Sys:

1. Critical requirement similarities: a comparison between the critical requirements collected from external sources and the concepts in the lexicon ontology;
2. Relationship analysis: a comparison between the relations of concepts in external sources and the relations present in the lexicon ontology;
3. Concept relationship: an analysis of concepts from external sources, regarding their similarity to the lexicon ontology.

In the fourth phase, "Knowledge Enrichment" (APKE4), the ontology that contains the translated knowledge is compared, through ontology mapping in the APKE-Sys, to lightweight ontologies that represent the domains (such as product, design, manufacturing) by a conceptual perspective. Those lightweight ontologies are related to consolidated models that represent their respective knowledge. The results of this phase are the ontologies that compose the Reference View of the IPDMS model, as seen on [23]. Lastly, the "Application in IPDP" (APKE5) phase comprises the addition of semantic rules to the reference ontologies and further application in the IPDP through the IPDMS. In this phase, a consistency analysis of the ontologies is done through an inference engine before and after the creation of the semantic rules, to check for inconsistencies in the mapping processes. The extraction, formalisation, and translation of knowledge to standardised representation can improve the implementation of the IPDMS and, consequently, the IPDP.

6 Experimental Case

6.1 Problem in Industrial Scenario

The application of the proposed approach was carried out in a Brazilian electronics manufacturer, here referred to as Company X. The company had issues related to poor communication in product development, and a few of its products had a high return rate. The company is currently implementing the IPDMS to coordinate its product development and wanted to use a more automated approach to extract product knowledge coming from customer-related data. In this case, which is experimental in nature, the chosen product was a 20 kVA Uninterrupted Power Supply (UPS). This version of the product took around 20 months of development, using more than 2800 h of work and costing around US$ 33.00000 (approximately). During the 20 months of development, the project entered in a 6-month hiatus due to reviews that were necessary, in order to the project be in attendance with needs from customers—this hiatus costs around US$ 13.43200 (approximately) to the enterprise. Furthermore, mistakes in the design caused malfunction while in use, after its launch, bringing more than 70% of products back to the manufacturer.

6.2 The Approach Application

The development process was brought back for its early stages for a full revaluation of its requirements and serves as the case for application of the proposed approach. The application is related to the early stages of PDP, more specifically in the requirements gathering and conversion into product features. The application is outlined in Fig. 3.

Firstly, in the "Knowledge Gathering" phase, the QFD of the new design of the UPS is collected in a "**.csv**" **format** and stored in a folder, accessed by a computational system, here called Approach for Product Knowledge Extraction System (APKE-Sys), that orchestrates the approach application (Detail A of Fig. 3).

In the "Knowledge Pooling" phase, the "**.csv**" **from QFD** was analysed by a **pattern analysis** algorithm in the APKE-Sys, in order to generate its tags and further structure the knowledge from the QFD in an ".xml" extension file (Detail B of Fig. 3). The ".csv" is analysed for keywords related to product, manufacturing, and design parameters. The identified patterns are put in "**.xml**" **tags**, later being

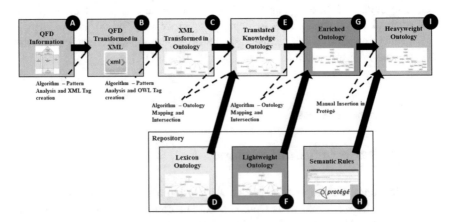

Fig. 3 Application of the proposed approach

joined and forming a file that contains the extracted knowledge from the QFD (Detail B of Fig. 3). This process basically organises the client's requirements to a hierarchical structure (i.e. a "change in the circuit board position" is converted into "Design → UPS → 20 kVA → Circuit Board" in the hierarchical structure).

For the "Knowledge Translation", the tags of the XML file were analysed by the APKE-Sys and **converted** into **".owl" tags**, creating an ontology based on the product knowledge extracted from the QFD (Detail C of Fig. 3). The basic structures of the ".owl" tags are kept in the system and are applied by an algorithm based on **pattern analysis**. Sequentially, the created "QFD Ontology" undergoes an **ontology mapping** process with the lexicon ontology (Detail D of Fig. 3). The commonalities between both ontologies are mapped, and an intersection of the ontology occurs, generating the "Translated Knowledge Ontology" (Detail E of Fig. 3). This phase merely creates an ontological structure of the XML file in accordance with company's specific nomenclatures (i.e. "20 kvA" is renamed as company's code "123ABC").

The "Knowledge Enrichment" process was, as his predecessor, a discrete process in the APKE-Sys, mapping the "Translated Knowledge" Ontology and the chosen domain (product, design, manufacturing) Lightweight Ontology (Detail F of Fig. 3) and combining both in an "Enriched" ontology (Detail G of Fig. 3). Like the process of the "Knowledge Translation", an algorithm for **ontology mapping** looks for similarities in both ontologies (similar classes, attributes and relations), adding the complementary information from the "Lightweight Ontology". This phase gathers the knowledge from the "Translated Knowledge Ontology" and distributes to their specific domain.

The "Application on IPDP" process was performed partially by the APKE-Sys and partially by the IPDMS. The Reference View, an input for application of IPDMS, was generated at the end of the "Knowledge Enrichment" process, by the addition **of semantic rules and consistency analysis in the Protégé software** (Detail H of Fig. 3) that define constraints. The final version of the ontologies offers improved semantics and improved quality on product knowledge. Those are Heavyweight Ontologies (Detail I of Fig. 3) that were validated by a team of specialists, in order to check their overall consistency with company nomenclature.

Firstly, in terms of time saving, the approach reduced an approximate total of 3 h of work from three professionals (9 h total in terms of cost) into a 15 min activity from one single professional (not counting the creation of semantic rules). This is translated into an improvement in time efficiency of more than 97% of previous development, while increasing product's quality with improved communication and semantic correctness, reducing design flaws. This reduction occurs through inconsistencies found on the reasoning process in the Heavyweight Ontology after the application in IPDMS. In terms of operational costs, the final cost of this process was reduced to US$ 292, representing a reduction of more than 97%.

7   Conclusion

The proposed method was able to standardise and extract the product's requirements, reducing the time and the cost of the project without reducing the quality of the final product in many aspects. This standardisation enabled the integration of product and manufacturing and presented reduced misinterpretations in product development.

This research provided a formalisation for the process of capturing information and improved communication and information sharing. The provided method approaches the three main identified issues in the literature review. The consistency analysis of information based on ontology mapping through the phases of Knowledge Translation and Knowledge Enrichment is one of the main strengths of the model, avoiding semantic heterogeneity and human mistakes while structuring the product's requirements.

Next steps of the research will focus in an expansion of the approach, adding more features to different cases, as means to explore the approach and stress its limitations, refining it further.

Acknowledgements The researchers would like to thank the company where the approach was applied, the DAI Program of the Brazilian National Council of Scientific and Technological Development (CNPq), and the Pontifical Catholic University of Paraná, for the funding and structure.

References

1. Szejka, A. L., Júnior, O. C., Loures, E. R., Panetto, H., & Aubry, A. (2016). Proposal of a model-driven ontology for product development process interoperability and information sharing. In *IFIP International Conference on Product Lifecycle Management* (pp. 158–168).
2. Pereira, J. A., & Junior, O. C. (2014). Product development model oriented for the R&D projects of the Brazilian electricity sector. *Applied Mechanics and Materials*, 366–373.
3. Chungoora, N., Young, R. I., Gunendran, G., Palmer, C., Usman, Z., Anjum, N. A., Cutting-Decelle, A. F., Harding, J. A., & Case, K. (2013). A model-driven ontology approach for manufacturing system interoperability and knowledge sharing. *Computers in Industry, 64*(4), 392–401.
4. Andreasen, M. M., & Hein, L. (1987). *Integrated product development*. IFS Publications Ltd.
5. Sosa, M. E., Eppinger, S. D., & Rowles, C. M. (2004). The misalignment of product architecture and organizational structure in complex product development. *Management Science, 50*(12), 1674–1689.
6. Pereira, J. A.. & Junior, O. C. (2014). Product development model oriented for the R&D projects of the Brazilian electricity sector. *Applied Mechanics and Materials, 518*, 366–373.
7. Panetto, H., Dassisti, M., & Tursi, A. (2012). ONTO-PDM: product-driven ontology for product data management interoperability within manufacturing process environment. *Advanced Engineering Informatics, 26*(2), 334–348.
8. Chungoora, N., & Young, R. I. M. (2011). Semantic reconciliation across design and manufacturing knowledge models: A logic-based approach. *Applied Ontology, 6*(4), 295–315.
9. Imran, M., & Young, B. (2013). The application of common logic based formal ontologies to assembly knowledge sharing. *Journal of Intelligent Manufacturing, 26*(1), 139–158.

10. Palmer, C., Urwin, E. N., Pinazo-Sánchez, J. M., Cid, F. S., Rodríguez, E. P., Pajkovska-Goceva, S., & Young, R. I. M. (2016). Reference ontologies to support the development of global production network systems. *Computers in Industry, 77*, 48–60.

11. Fensel, D. (2004). Triple-space computing: Semantic web services based on persistent publication of information. In *Intelligence in communication systems* (pp. 43–53).

12. Fahad, M., Moalla, N., Bouras, A., Qadir, M. A., & Farukh, M. (2010). Disjoint-knowledge analysis and preservation in ontology merging process. In *Software engineering advances (ICSEA)* (pp. 422–428).

13. Junior, O. C., & Young, R. I. M. (2003). Information sharing in multiviewpoint injection moulding design and manufacturing. *International Journal of Production Research, 41*(7), 1565–1586.

14. Haveman, S. P., & Bonnema, G. M. (2013). Requirements for high-level models supporting design space exploration in model-based systems engineering. *Procedia Computer Science, 16*, 293–302.

15. Nattermann, R., & Anderl, R. (2010). Approach for a data-management-system and a proceeding-model for the development of adaptronic systems. In *ASME 2010 International Mechanical Engineering Congress and Exposition* (pp. 379–387).

16. Moneva, H., Hamberg, R., & Punter, T. (2011). A design framework for model-based development of complex systems. In *32nd IEEE Real-Time Systems Symposium 2nd Analytical Virtual Integration of Cyber-Physical Systems Workshop*, Vienna.

17. Junior, O. C., & Young, R. I. M. (2010). Information mapping across injection moulding design and manufacture domains. *International Journal of Production Research, 48*(15), 4437–4462.

18. Stechert, C., & Franke, H. J. (2009). Managing requirements as the core of multi-disciplinary product development. *CIRP Journal of Manufacturing Science and Technology, 1*(3), 153–158.

19. Liao, Y., Lezoche, M., Panetto, H., Boudjlida, N., & Loures, E. R. (2015). Semantic annotation for knowledge explicitation in a product lifecycle management context: A survey. *Computers in Industry, 71*, 24–34.

20. Szejka, A. L., Aubry, A., Panetto, H., Junior, O. C., & Loures, E. R. (2014, October). Towards a conceptual framework for requirements interoperability in complex systems engineering. In *OTM Confederated International Conferences* (pp. 229–240).

21. Szejka, A. L., & Junior, O. C. (2017). The application of reference ontologies for semantic interoperability in an integrated product development process in smart factories. *Procedia Manufacturing, 11*, 1375–1384.

22. Cleland-Huang, J., Chang, C. K., Sethi, G., Javvaji, K., Hu, H., & Xia, J. (2002). Automating speculative queries through event-based requirements traceability. In *Requirements engineering* (pp. 289–296).

23. Szejka, A. L., Leite, A. F. C. S. M., Canciglieri, M. B., & Junior, O. C. (2016). Structuring a foundation basis for semantic interoperability in product development process. In *ISPE TE* (pp. 957–966).

Towards Adaptive, Interactive, Assistive and Collaborative Assembly Workplaces Through Semantic Technologies

Izaskun Fernandez, Patricia Casla, Iker Esnaola, Laure Parigot, Angelo Marguglio, and Teegan Johnson

Abstract Assembly systems are characterised by being mainly manual labour environments with high flexibility but low productivity. To increase productivity while maintaining flexibility, assembly systems need to be redesigned by incorporating automation mechanisms and assistance tools that adapt themselves to the context and complement human capabilities. In this paper, we present a semantic approach which can adapt the workplace in real time to the production context and operators' characteristics. The approach is based on a semantic representation of the workplaces, processes and workers' profiles, as well as their environmental situation, like a workplace digital twin. Furthermore, the approach guides operators in a personalised way providing intuitive communication channels such as voice and gestures to interact with the automatisms in place, ensuring the process execution correctness and operators' satisfaction. The approach is validated in two specific assembly workplaces, demonstrating the easy adoption of it in different scenarios.

I. Fernandez (✉) · P. Casla · I. Esnaola
TEKNIKER, Basque Research and Technology Alliance (BRTA), Iñaki Goenaga 5, 20600 Eibar, Spain
e-mail: izaskun.fernandez@tekniker.es

P. Casla
e-mail: patricia.casla@tekniker.es

I. Esnaola
e-mail: iker.esnaola@tekniker.es

L. Parigot
Airbus Group, Route de Bayonne 316, 31060 Toulouse, France
e-mail: laure.parigot@airbus.com

A. Marguglio
Engineering—Ingegneria Informatica S.P.A., Via San Martino Della Battaglia 56, 00185 Roma, Italy
e-mail: angelo.marguglio@eng.it

T. Johnson
IPHF Group, Cranfield University, Bedfordshire MK43 0AL, UK
e-mail: t.l.johnson@cranfield.ac.uk

Keywords Adaptive · Interactive · Assistive and collaborative workplaces ·
Manufacturing ontology · Manufacturing agent's interoperability

1 Introduction

Sectors characterised by small batch production and complex products (e.g. aeronautics) need to combine high levels of flexibility with high productivity rates. In such sectors, assembly and auxiliary operations are mainly performed by humans as they bring inimitable agility to adjust to changes, as well as skills that cannot be replaced by automation. However, manual intensive activities can also present disadvantages such as potential physical or mental limitations that can restrict overall performance of the assembly system.

In a scenario with ever-changing demands, assembly systems need to put together humans and automation taking advantage of each other's strengths to balance flexibility and productivity requirements in an easy and cost-effective way. This collaboration raises challenges that must be faced to get a successful collaborative workplace: human and robot must know about each other situation; they must be able to interact naturally; and personalised and adapted support must be provided to operators specially in new assembly processes.

In this paper, we present a generic approach for new assembly scenarios that face all these challenges based on a semantic approach. Section 2 presents the related work. Section 3 describes the semantic approach including the main components overview and a detailed description of the VAR ontology. Section 4 presents the application of the system with the corresponding ontology instantiation of two use cases. Section 5 includes the discussion of both experiences. Finally, in Sect. 6, conclusions and future work are presented.

2 Related Work

One of the main issues related to the presence of the Semantic Technologies in the manufacturing domain is the lack of generally accepted and available ontologies. Furthermore, although some proposals have been done during recent years, few of them are public and available for reuse.

On the one hand, there are ontologies aimed at covering the manufacturing domain area, such as Manufacturing's Semantic Ontology (MASON) [1], the P-PSO Ontology [2] or the (Manufacturing Core Concepts Ontology (MCCO) [3]. On the other, there are ontologies covering a very specific area of the manufacturing domain. ExtruOnt ontology [4] aims at describing an extruder, CM-Core ontology [5] is aimed at representing the core entities of the condition, PRONTO Ontology (Product Ontology) [6] captures the core concepts to represent products, Ontology of Standard of the Exchange of Product model data (OntoSTEP) [7] aims at representing product information but focusing on their geometry, and the Manufacturing Service Description Language (MSDL) [8] ontology represents the production service capabilities. However, none of the mentioned ontologies deal with the information exchange required among the different agents in manufacturing scenarios. Although they do not ensure interoperability in a semantic level, there is a group of relevant and extended standards that have been developed for information exchange in the manufacturing domain.

Business To Manufacturing Markup Language [9] (B2MML) is an XML implementation of IEC/ISO 62264 that is an international standard for enterprise-control system integration. B2MML is meant to provide a common data definition to link enterprise resource planning (ERP) and supply chain management (SCM) systems with manufacturing systems such as Industrial Control Systems (ICS) and Manufacturing Execution Systems (MES).

AutomationML [10] aims to standardise data exchange in the engineering process of production systems. Therefore, AutomationML e.V. develops and maintains an open, neutral, XML-based, and free industry data representation standard which enables a domain and company crossing transfer of engineering data.

eCl@ss [11] has established itself internationally as the only ISO/IEC-compliant industry standard and is thus the reference data standard for the classification and unambiguous description of products and services. With the help of eCl@ss, standardised digital data transfer is enabled. As a result, classifications and product description properties can be exchanged across the value chain.

3   Semantic-Oriented Framework

The semantic-oriented framework aims to support adaptive, interactive, assistive and collaborative assembly workplaces in an ever-changing scenario by providing: (1) plug- and-produce mechanisms to enable the reconfiguration of the workplaces; (2) natural communication enhancing human-automatism collaboration, (3) adaptation of the workplaces to the dynamic conditions of the environment and (4) personalised, context-aware guidance in the execution of productive tasks [12]. Figure 1 shows the set of key generic components of the framework and implements the aforementioned mechanisms by exploiting and exchanging the information through a central semantic repository based on a core ontology named VAR.

The green components (*Mediation Services*, *Device Manager* and the *Multimodal, Multichannel Interaction Manager*) collect real-time context information from operators, automatisms (i.e. such as robots, machines or smart tools) even legacy systems and executing adaptation commands. While the blue components (*Event Manager, Collaborative Asset Manager and Semantic Repository*) enable real-time adaptation as well as personalisation, and finally, the orange ones (*Decision Support System, Collaborative Knowledge Management* and the *VR/AR-Based Training and Guidance*) are the ones in charge of providing context-aware assistance.

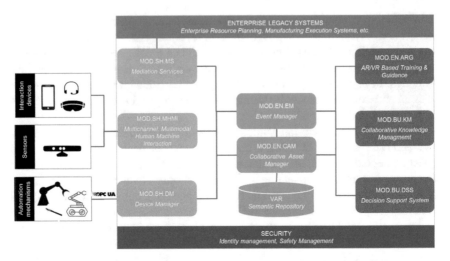

Fig. 1 Semantic-oriented framework reference implementation for collaborative workplaces

The *Mediation Services, Device Manager* and the *Multimodal, Multichannel Interaction Manager* components manage and interact with all the agents in the workplaces (i.e. legacy systems, automatisms and operators) and gather all the relevant real-time information coming from them. In particular, the *Mediation Services* enables collection of dynamic information about the operator involved and the operation in progress from Manufacturing Execution Systems; the *Device Manager* supports automatisms discovery by identifying the methods and variables exposed as well as status update (e.g. regarding the automatism itself or the assembly process), and the *Multimodal Interaction Manager* manages the commands coming from operators. All these components must verify the exchanged information and, if it is correct, include it in the semantic repository through the CAM component.

Then the *Event Manager* triggers the adaption and notification commands based on the defined rules and the dynamic context information stored in the semantic repository. Finally, the *Decision Support System, Collaborative Knowledge Management* and the *VR/AR-Based Training and Guidance* components consume the commands and notifications triggered by the *Event Manager* and all the information gathered in the semantic repository that reflects the dynamic and realistic view of the manufacturing process. Furthermore, they aim to assist operators the best way possible, considering their profiles and the dynamic context.

Operators are provided with the required knowledge and process definition and dynamic status information through the semantic repository according to the VAR ontology. This enables the reusability of all the components in different scenarios without any modification exception for the semantic repository, which requires a new instantiation of the VAR ontology for each scenario according to the targeted assembly process.

3.1 VAR Ontology

The VAR ontology is the core element in the semantic-oriented framework, enabling the data exchange from and to diverse agents in the assembly scenarios including external sources such as legacy systems (e.g. Manufacturing Execution Systems), operators, robots, tools and so on to make possible adaptive, interactive, assistive and collaborative assembly workplaces.

The VAR ontology was developed following the well-known NeOn methodology [13]. First, a group of manufacturing experts defined the scenario requirements, which were later registered in the form of Competency Questions (CQs) in the Ontology Requirements Specification Document (ORSD). These requirements included adaptation to workplace environmental conditions, natural interaction between humans and machines and optimal automation configuration among others. From these CQs, the main ontology concepts were extracted.

The VAR ontology's design has been based on the B2MML standard in order to enhance interoperability with external legacy systems such as ERP and MES. In the context of the SatisFactory[1] project, this standard was translated into OWL. Following the ontology reuse best practices, a total of 18 classes and 48 properties have been reused by the VAR ontology. As for the requirements of the new assembly workplaces which were not covered in the B2MLL OWL version, a set of new resources were defined in the VAR ontology. As a result, the VAR ontology is composed by 86 classes, 97 object properties and over 70 data properties.

The VAR ontology follows a modular approach avoiding strong dependencies between modules in order to empower its module's reuse, to support more efficient query answering and to enhance modules' evolution [14]. Furthermore, this ontology modularisation has been undertaken from the ontology design stage to avoid performing arduous and time-consuming ontology modularisation techniques in the future.

It is worth mentioning that the VAR ontology does not contain any contradictory facts, as a Pellet reasoner has shown its logical consistency. This consistency feature is of utmost importance for the VAR ontology, as autonomous software agents may perform reasoning tasks with instantiations and come to conclusions without human supervision. Therefore, without ensuring ontology consistency, wrong conclusions could be deduced. Additionally, all the defined CQs are adequately addressed by the VAR ontology; thus, it is considered verified.

The ontology can be divided into four main modules: manufacturing assets; plug and produce; traceability and interaction. The modules are related to each other through five main properties connecting classes from different modules, as it is shown in Fig. 2.

---

[1] http://www.satisfactory-project.eu/satisfactory/.

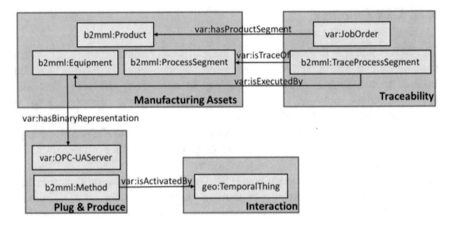

Fig. 2 VAR ontology main modules

Manufacturing Assets Module

The manufacturing assets module contains all the relevant classes and properties for defining products produced by assembly processes. It includes physical entities (tangible assets) in assembly workplaces, such as product, material, equipment, personnel and interaction devices, as well as non-physical entities (intangible assets) like processes. A UML representation of the excerpt of the VAR ontology of the manufacturing assets is shown in Fig. 3.

The product is represented by the *ProductDefinition* class that is composed of product segments (*ProductSegment* class). In turn, each product segment can be made of a set of product segments following a dependency flow (*hasSegmentDependency*). Each product segment is defined by a process segment (*ProcessSegment* class) that represents the personnel and equipment resources required to carry out a production step, and it can be made of a set of process segments following a dependency flow.

A person (*Person class*) represents a specifically identified individual with each own characteristics and capabilities and can be described by a set of properties

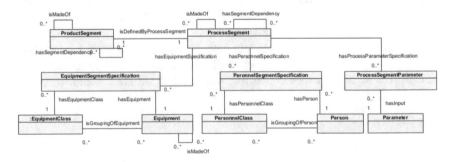

Fig. 3 UML diagram of tangible assets representation excerpt in VAR ontology

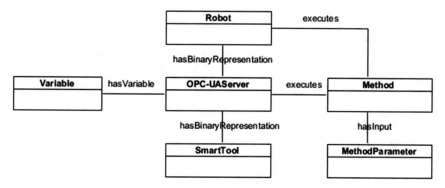

Fig. 4 UML diagram of plug and produce excerpt in VAR Ontology

(*PersonProperty* class) that can be grouped based on the *personPropertyType* data property in a specific *PersonnelClass*.

Plug and Produce Module

The plug and produce module defines all the necessary classes and properties for supporting automatisation in assembly workplaces such as OPC-UA server, methods and variables that equipment provides.

Both robots (*Robot* class) and smart tools (*Smarttools* class) are automation mechanisms involving adaptation capabilities and are represented by an OPC-UA server (*OPC-UAServer* class) to support the standard-based plug-and-produce approach. The *OPC-UAServer* class provides a binary representation of both, and it monitors variables (*Variable* class) linked to changes in robot/smart tool properties and execute methods (*Method* class). Furthermore, the methods can involve a set of parameters (*MethodParameter* class) as shown in Fig. 4.

Traceability Module

The aim of the traceability module is to gather all the necessary trace information. For that, it includes classes like JobOrder or TraceProcessSegment. Furthermore, this module enables to have in real time actual context status control to support adaptation capabilities. The involved object properties are updated in run-time according to the real situation: some of them directly through the services provided by the CAM, and others based on semantic rules, property chains and logical inferences. The current situation is controlled by the job order (*JobOrder* class) which is linked to the operation (*ProcessSegment* class) in progress as well as the involved equipment (Equipment class) and worker(s) (*Person* Class) through the *TraceProcessSegment* and its links to the rest of the instances of the mentioned classes (like *JobOrder isExecutedBy Equipment*, or *Person isLoggedIn Equipment*.

Interaction Module

For interaction issues, the ontology includes individuals like *Start*, *Stop*, *Resume* and *Move* belonging to the class *BasicAction* (a subclass of *TemporalThing* class) that are used to determine the commands that can be used to interact with the automatisms linked to their *Methods*.

This module is also in charge of representing the notifications and the related channels as well as the interaction devices that supports the natural and adapted interaction through *Notification*, *Channel* and *InteractionDevice* classes, respectively.

4 Use Cases

To demonstrate the easy adoption of the semantic approach presented in the previous section, we have deployed such system in two assembly scenarios: (1) optimisation of the assembly and tightening of the hydraulic system on the A350 over wing panel (OWP) including automatic tool configuration, on the job guidance and traceability at Airbus and (2) the collaborative assembly of a latch valve where the system adapts itself to the operator's characteristics and the operator interacts with the Manufacturing Execution System, an industrial assembly robot and a mobile logistic robot in a natural way (i.e. using voice and/or gestures) at Tekniker's facilities.

In both scenarios, the VAR ontology was instantiated detailing the corresponding process step by step: including all the task dependency restrictions as well as all the parametric configurations. Furthermore, during task execution, once the corresponding automatism (smart torque wrench, dual arm robot or logistic robot, respectively) was discovered, the related OPC-UA servers is, automatically, instantiated. Figure 5 includes an RDF excerpt that shows, in the Airbus scenario, the smart tool *M05* discovered and related to the specific work centre (*MS40.A*) where it is operative. The OPC-UA and smart tool information is provided by the smart tool itself once it is discovered, publishing it in the semantic repository through the CAM services. As for the IP and the work centre, they are dynamically updated when discovered.

```
var-tek:M05 rdf:type :SmartTool;

        ...

        :hasBinaryRepresentation :stOPCUServerM05 .
:stOPCUServer1731M005 rdf:type :OPC-UAServer;
        :ip "107.18.31.5" ;
        :hasVariable :MalfunctionFallDetection1731M005;

        ...

        :executes :setWrenchMode1731M005.
:MS40.A rdf:type b2mml:Equipment ;

        ...

        :ipRangeEthernet "107.18.31.00;107.18.31.20" ;
        :isMadeOf var-tek:M05 .
```

Fig. 5 Smart tool discovery RDF excerpt

```
:soi.001-Task_03_04_01-JO_soi001 rdf:type :TraceProcess-
Segment;
    :finished "1" ;
    :endTimestamp "2019-07-17T08:32:48Z" ;
    :personId "1" ;
    :actualParameterValue "23.56" ;

    ...

        equipmentId "M05" .
var-tek:M05-07-17T08:41:28Z rdf:type :EquipmentEvent;
    :equipmentId "M05" ;
    :timestamp "2019-07-17T08:41:28Z" ;
    :eventType "STATUS" :
```

Fig. 6 Dynamic status RDF excerpt

In addition, during the operation, the dynamic current status, including full traceability, is updated. For instance, in Fig. 6 you can see the *soi.001-Task_03_04_01-JO_soi001* operation trace individual with the real reached workbench value (*23.56*), the operation start and end time as well as who has participated in such an operation.

The RDF excerpt also includes a malfunction reported by the smart tool (*M05-07-17T08:41:28Z*) during the job order execution. All this information is exploited by the quality and metrology personnel to supervise the task execution, identify potential conflictive operations and even decide on the life of certain smart tool.

5 Results Discussion

The Airbus's evaluation involved seven participants completing an experiment where they were trained to use the HoloLens and smart tool on a mock-up before completing a hydraulic pipe installation in the OWP of a test aircraft. The usability was explored through surveys gathering quantitative data on usability and mental workload. The results showed a good level of usability for all usability dimensions. The usability and mental workload mock-up scores were better than the scores obtained from the participants after completing the task on the OWP. Another potential benefit identified is the improvement of the productivity due to the reduction of time required to search for information and to change tool or the increased traceability as everything can be recorded and reported.

In Tekniker's evaluation involved, twenty participants completed the assembly process and included the assessment of usability (i.e. including both the gesture and voice-based interaction), mental workload and trust in human–robot interaction. The usability, mental workload and trust scores were all positive, indicating good usability for the system. The participants' responses indicated that they found the voice input more usable than the gesture inputs, and this may have resulted from the ability to use their natural language rather than having to remember the gestures to use. Furthermore, some potential benefits such as an increase in productivity due to a reduction of the displacements have been identified.

6 Conclusion and Future Work

This paper tackles the challenges that arise when putting humans and automation together in collaborative manufacturing scenarios, to leverage each other's strengths to balance flexibility and productivity requirements in an easy and cost-effective way. Towards that goal, a generic semantic-oriented framework based on the VAR ontology has been developed, including modules addressing: (1) automatisms plug-and-produce mechanisms to enable dynamic reconfiguration of the workplaces; (2) natural communication enhancing human-automatism collaboration; (3) adaptation of the workplaces to the dynamic conditions of the environment and (4) personalised, context-aware guidance in the execution of productive tasks. All these modules take advantage of the real- time semantic information representation, according to the VAR ontology.

The reusability of the generic approach has been demonstrated by deploying the framework in two real scenarios, and the experimentations carried out in them show that the functionalities supported by the framework are well accepted and exploited by the users, leading to an increase in productivity even in changing environments.

Acknowledgements This work is based on the results of the A4BLUE project which has received funding from the European Union's Horizon 2020 research and innovation programme under grant agreement no. 723828. It is worthy also to thanks to École Polytechnique Fédérale de Lausanne (EPFL) for providing the B2MML OWL version, as well as to IPHF Group from Cranfield University for leading and ensuring the correct management of the ethical aspects needed in the study.

Statement of Ethical Approval
This research was approved (CURES/7646/2019) by the Cranfield University Research Ethics Committee (CURES). It was conducted in accordance with the Cranfield Research Integrity Policy, the British Psychological Society's Code of Human Research Ethics [15] and the General Data Protection Regulation 2018. To comply with this, written informed consent was obtained from participants after they had been made aware of the nature of the study. The informed consent form was provided to the participants along with a briefing sheet describing the nature of the experiment, objectives, procedure and timing. The right to withdraw, including the timeframe for withdrawal, was included within both the briefing and consent forms.

Only a numeric identifier was stored in the specific instantiation of the ontology so no personal data enabling the identification of the participants were shared.

References

1. Lemaignan, S., Siadat, A., Dantan, J. Y., & Semenenko, A. (2006, June). MASON: A proposal for an ontology of manufacturing domain. In *IEEE Workshop on Distributed Intelligent Systems: Collective Intelligence and Its Applications (DIS'06)* (pp. 195–200). IEEE.
2. Garetti, M., & Fumagalli, L. (2012). P-PSO ontology for manufacturing systems. *IFAC Proceedings Volumes, 45*(6), 449–456.
3. Usman, Z., Young, R. I. M., Chungoora, N., Palmer, C., Case, K., & Harding, J. (2011, March). A manufacturing core concepts ontology for product lifecycle interoperability. In *International IFIP Working Conference on Enterprise Interoperability* (pp. 5–18). Springer, Berlin, Heidelberg.
4. Ramírez-Durán, V. J., Berges, I., & Illarramendi, A. ExtruOnt: An ontology for describing a type of manufacturing machine for industry 4.0 systems.
5. Cao, Q., Zanni-Merk, C., & Reich, C. (2019). Towards a core ontology for condition monitoring. *Procedia Manufacturing, 28*, 177–182.
6. Vegetti, M., Leone, H., & Henning, G. (2011). PRONTO: An ontology for comprehensive and consistent representation of product information. *Engineering Applications of Artificial Intelligence, 24*(8), 1305–1327.
7. Krima, S., Barbau, R., Fiorentini, X., Sudarsan, R., & Sriram, R. D. (2009). *Ontostep: OWL-DL Ontology for Step.* National Institute of Standards and Technology, NISTIR, 7561.
8. Ameri, F., Urbanovsky, C., & McArthur, C. (2012, July). A systematic approach to developing ontologies for manufacturing service modeling. In *Proceedings of the Workshop on Ontology and Semantic Web for Manufacturing* (p. 14).
9. Business to Manufacturing Markup Language (B2MML) Site. http://www.mesa.org/en/B2M ML.asp. Last accessed November 15, 2019.
10. AutomationML Site. https://www.automationml.org/o.red.c/organisation.html. Last accessed November 15, 2019.
11. eclass Site. https://www.eclass.eu/en/standard/search-in-eclss.html. Last accessed November 15, 2019.
12. Fletcher, S. R., Johnson, T., Adlon, T., Larreina, J., Casla, P., Parigot, L., Alfaro, P. J., & Otero, M. (2019). Adaptive automation assembly: Identifying system requirements for technical efficiency and worker satisfaction. *Computers & Industrial Engineering, 0360-8352,* 105772.
13. Suárez-Figueroa, M. C., Gómez-Pérez, A., & Fernandez-Lopez, M. (2015). The NeOn methodology framework: A scenario-based methodology for ontology development. *Applied Ontology, 10*(2), 107 145.
14. Grau, B. C., Horrocks, I., Kazakov, Y., & Sattler, U. (2008). Modular reuse of ontologies: Theory and practice. *Journal of Artificial Intelligence Research, 31*(2008), 273–318. https://doi.org/10.1613/jair.2375
15. The British Psychological Society. (2014). Code of Human Research Ethics. https://www.bps.org.uk/sites/beta.bps.org.uk/files/Policy/Policy%20-%20Files/BPS%20Code%20of%20H uman%20Research%20Ethics.pdf. Retrieved February 01, 2019

A Semantic Interface Model to Support the Integration of Drones in a Cyber-Physical Factory

S. A. Puviyarasu⬥, Farouk Belkadi⬥, Catherine da Cunha⬥, Abdelhamid Chriette⬥, and Alain Bernard⬥

Abstract Industries rely on digital transformation and emergent technologies in order to reach processes efficiency and flexibility. Cyber-physical systems are currently proposed as an answer to this fourth industrial revolution. In this context, the use of drones can offer new opportunities in manufacturing shop floors. However, the adoption of such technologies requires the development of integrated interfaces to solve interoperability issues between the different components of the global production system. This paper addresses the problem of interfacing through a conceptual interface model representing the exchanged data flows between drones and a given production system. It is instantiated on a real experimental 4.0 platform combing production modules and drones. This model aims to support communication between the related logical and software components. This work is a required step in order to later address the interoperability issues.

Keywords Cyber-physical system (CPS) · Interface model · Production system · Drones

1 Introduction and Problem Statement

Industry 4.0 is mainly about new innovative business models embracing the new possibilities offered by new technologies. Cyber-physical systems (CPS) are currently proposed as an answer to the new industrial revolution. The main characteristic of CPS is a coupling of physical and cyber components with a networked connection [1]. It involves computational elements able to operate on different scales and interact with each other when a change in context. The CPS technology and application for production is termed, as cyber-physical production system (CPPS) for the last few years [2]. They offer an effective and productive system by enabling smart manufacturing. The CPPS controls and integrates with various smart devices and information systems with standard interfaces [3]. Incorporating drones into industry offers

S. A. Puviyarasu (✉) · F. Belkadi · C. da Cunha · A. Chriette · A. Bernard
Ecole Centrale de Nantes, LS2N-UMR 6004 Nantes, France
e-mail: Puviyarasu.sa@ls2n.fr

new reliable industrial applications, not only transportation but also for inspection and data gathering activities [4].

A real smart platform 4.0 is used as a case for integrating the devices with existing systems. The cyber-physical factory (CP Factory) includes advanced technologies with networked connected modules [5]. In this context, we incorporated a new drone system for material transportation applications. However, adopting such technologies with the existing complex system raises interface issues. In particular, interfacing problems between the different components of the global production system requires to develop an integrated approach. These issues need to be addressed before tackling interoperability issues. The model-driven architecture (MDA) approach will be followed. It is traditionally used to identify the integration or interoperability of the system from the conceptual description and to the implementation [6]. MDA is a set of guidelines for structuring the specifications and issues at each abstraction level of system integration. They are expressed in models to facilitate the trans-level of systems. This paper describes a conceptual interface model to represent communication between the system's logical component and software level. The complete collaborative existing and proposed interface model of a CP factory is shown. The proposed conceptual semantic interface model is the first step to future physical implementation.

Section 2 describes the state of the art of current issues and ongoing developments of drone integration. In Sect. 3, a short description of the system architecture and analysis is proposed. The proposed semantic interface model at logical and software levels is discussed in Sect. 4. Finally, conclusions are drawn and future works identified.

2   State of the Art

Industry 4.0 combines the virtual and physical world of production, machines, systems, and sensors to communicate with each other, to share information, and control each other independently [5]. Cyber-physical systems (CPS) are systems in which the collaborating physical and software components are deeply intertwined, able to operate and interact with each other were change in context. The cyber-physical factory (CP factory) has a highly complex and digital production that provides the smart factory platform. The CPS applications for production integrate state-of-the-art technologies and reinforce technology interactions [7]. Flying drones are increasingly popular for the industrial context and it offers new possibilities for transforming industries. Drones can operate without any human intervention for prolonged periods and can recognize with a full 360° overview industrial objective [8]. They are becoming popular intelligent logistics tools in manufacturing and other industries. For instance, Amazon introduced the prime air for the commercial delivery of shipping products to customers [9]. Load transportation and deployment by drones are handy for many applications including the delivery of packages, delivery of isolated victims in disasters (floods, earthquakes, fires, industrial disasters, and many others. It is also a fundamental technology for other future applications [10]. With increased technological capabilities and connectivity, drones are one of the latest technologies to fit into the industry, at various stages.

However, incorporating a drone into a CPPS is a challenging task in terms of interoperability [11]. These technologies and application for the industry would first require the installation and supporting units defining the interoperable connectivity that fit into another system [12]. Drones and their interoperability represent one of the most important areas of innovation across the technology industry [13]. To tackle the interface and interoperability problem at each abstraction level, the model-driven architecture (MDA) helps to solve the issue of the system. The use of this approach allows a complete follow-up from expressing system interface and interoperability requirements to solution coding [14]. Therefore, appropriate models matching the particular objectives should be used. Semantic approaches can help to reunite the objectives and facilitate interoperability irrespective of the domain. The semantic model is the process of interrelation information and systems from diverse sources and facilitate (or even automate) the communication between the software and components [15]. Below section shows and discussed the semantic interface of diverse source and facilitate the communication exchanges between the different modules in the CP factory. Next section describes the system architecture and analysis.

3   System Architecture and Analysis

This section presents a short overview of the cyber-physical production system and flying drones deployed in the LS2N[1] laboratory. Figure 1 shows the targeted future scenario in the CP factory with a drone. Each system has its own architecture and functionality to perform its allocated task. Only the initial description of each system is discussed below.

3.1   CP Factory

The CP factory reflects the Industry 4.0 production paradigm by offering a modular CPPS. The platform is based on a FESTO standard solution. The core of the system is its modularity, which enables great edibility by combining modules in different configurations for a variety of applications [5]. This system assembles mobile cases in a standard and customized way. The final product and its components are small and light: fuses, printed circuit board, bottom, and upper cover. The system includes various assembly stations, warehouses, and various application modules. The basic

---

[1] Laboratoire des Sciences du Numérique de Nantes, LS2N-UMR 6004, France.

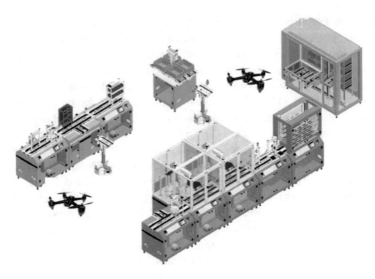

Fig. 1 Target scenario in the platform

module has different functionalities corresponding to the application module. The current CP factory uses humans and mobile robots for material transportation. An manufacturing execution system (MES) executes the operations. The other fleet-deploy processes the operation of the mobile robot by navigating the boundaries, data and signal processing. Those elements can work separately and a platform called "fleet manager" coordinates their interactions. The software components of these systems are deeply intertwined to support the global assembly process.

3.2 *Flying Drone*

The flying drone has a standard architecture composed of Raspberry Pi, Pixhawk, vision board camera, guidance system, proximity sensor, parallel motion 2-jaw gripper, suction gripper, and more. Generally, the drone software comprises various dimensions controlling platform, mapping and cloud-based platform [16]. The primary issue for the indoor application of the drone is the localization and mapping. Simultaneous localization and mapping (SLAM) approach is being developed to address these issues [17]. A more detailed description of the system is out of scope for this paper.

3.3　Integration Scenario

In this work, we want to use the drone for the transportation tasks done by humans in the current situation: fuses and top covers Magasin filling (cf. Fig. 1).

The feasibility analysis is done at the physical and also software level. The requirements of the possible use-case scenario for the various modules of the platform have been identified. The physical interfaces between drone and CP factory are "magazine application module" and "fuse application module." The software interfaces are the existing fleet coordination and smart soft components of a drone. The results of the feasibility analysis confirm that the target scenario can be implemented. However, it requires a semantic component level integration and the development of transfer units of the application module corresponding to drone physical interfaces.

4　Proposed Semantic Interface Model

The semantics framework serves many purposes, from structuring to solving the interoperability issues. It assists in identifying the scope to build shared data resources or information exchanges. The semantic model helps at each abstraction level to incorporate the new possibilities of integration of the system. The proposed semantic interface model is at two levels: logical component and software.

4.1　Semantic Interface Architecture-Logical Component Level

In this section, the semantic interfacing relationship between the system's logical component levels is proposed for the material transportation application. The proposition considers the actual ports of the existing system but only at the black-box level. From the drone system point of view, the coupling and decoupling of material for each application module are the same. However, each application module has a different type of physical transfer unit. It may affect the component level interface with a new device. Figures 2 and 3 show the conceptual logical component interface model proposed for the target system, modeled using SysML[2]. The interface model shows the common communication relationship between both systems in order to perform a task. The logical component and port of the system should have fine adjustments for docking-undocking the fuses. Laser retroreflectors (a technology already used in the CP Factory) will ensure the interoperable communication between the ports of components. Firstly, Fig. 2 shows the interface model of logical components between the fuse application module and the drone.

---

[2] https://sysml.org/.

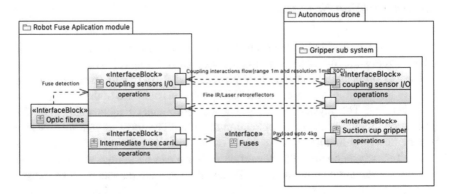

Fig. 2 Fuse application module-drone logical interface model

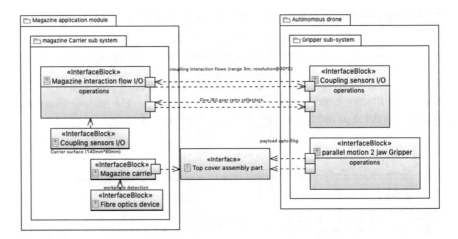

Fig. 3 Magazine application module-drone logical interface model

Secondly, Fig. 3 shows the interface model of logical components between the magazine application module and the drone.

4.2 Semantic Interface Architecture—Software Level

Section 3 summarizes the existing CP factory software components and the feasibility of the target scenario. This section specifies the interface architecture of the software level component to support the goal. The visual representations use UML's component diagram syntax.

4.2.1   Smart Soft Component of the Target System

The drone system has various dimensions in the software level, which includes end-user software, robot operating software, operating software, and a cloud-based control platform. The emerging smart soft is a component approach for drone software development and software-based communication patterns, which is a core for drone's component modeling [17]. Using these distributed components, the communication patterns predefine the semantics of the interface of components, irrespective of where they applied [18, 19]. For our target system, the smart soft approach is used on diverse components namely: guidance system, knowledge-base, SLAM (simultaneous localization and mapping) map server, base server, and docking-undocking mode, synchronize state. These components are predefined intelligence of drones. Figure 4 shows the conceptual semantic interface model of drone. This model defines interfacing components and their communication exchange between the different components. From the fig, it is shown that the smart soft components of the drone

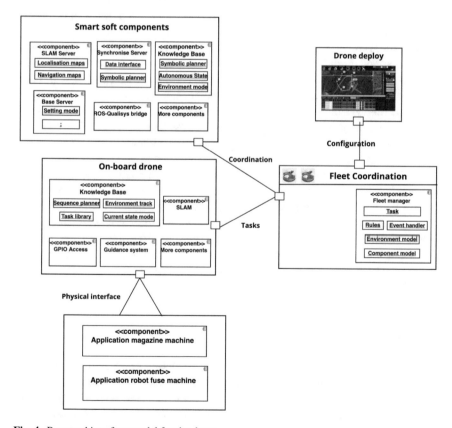

Fig. 4 Proposed interface model for the drone

are intertwined and coordinates with the fleet manager to exchange the predefine intelligence of drone.

Then drone deploys which is another application used to configure the drone and support the processing, mapping, and exchange of other dynamic information with the fleet manager. The collaborative semantic interface model and their communication exchanges for the global target scenario (Fig. 1) are presented in the following section.

4.2.2   Collaborative Semantic Interface Model of the Target System

The proposed collaborative semantic interface model gives a feasible and preliminary solution to integrate drones in the CP factory. This section shows the existing system software components and specifies the collaboration interactions with the proposed system semantic model at the software level.

Figure 5 shows the collaborative interface model at the software level of the target system. Manufacturing execution system (MES), other fleet deploys, fleet managers, and other fleets are the existing software components in CP factory. The drone deploys, smart soft components, onboard components are the drone software components. Three collaborative interactions between existing and drone software levels for the targeted scenario in CP factory are proposed. (1) Fleet manager performs the coordinative functionality of data processing, mapping, environment positioning, configuration of the system (drone deploys, other fleets, MES). For example, the drone deploys sends the drone initialized command to the fleet manager. The command helps the fleet manager position the drone for transportation. (2) The smart soft components of mobile robots and drones are deeply intertwined with fleet coordination. For example, the task library of the fleet manager receives and synchronize the predefined intelligence of drone for docking and undocking the materials. (3) In order to perform a global task of the production process, the interoperable information exchanges of (1) and (2) shares the configuration information, predefined intelligence of the system, for example, synchronization of production tasks with drone task. These three collaborative communication exchanges at the software level for the targeted scenario are necessary to perform reliable drone transportation.

5   Conclusion and Future Work

This paper addresses the integration of drones in a cyber-physical factory, as a starting step to create new possibilities for reliable transportations. This paper describes the semantic interface model and the interactions between elements of the system. In this work, we identified the interfacing components and communication exchanges between the drone and the production system. Then, we proposed the semantic interface model on the logical component and software application levels using MDA. From the identified communication exchange, interface requirements are identified.

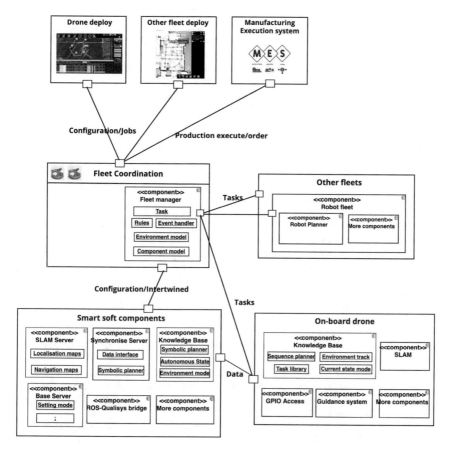

Fig. 5 Proposed collaborative semantic interface model for the global target system

It is the required step to analyze and address the interoperability problems further in future work. Finally, the design and realization of the interoperability between the different components at physical (e.g., coupling sensors) and software (e.g., synchronization, drone task planning) level of the system will be achieved.

References

1. Greer, C., Burns, M., Wollman, D., & Griffor, E. *Cyber-physical systems and Internet of Things.* National Institute of Standards and Technology, Special Publication 1900-202
2. Cardin, O. (2019). Classification of cyber-physical production systems applications: Proposition of an analysis framework. *Computers in Industry, 104,* 11–21. Elsevier, https://doi.org/10.1016/j.compind.2018.10.002.
3. Capitanelli, A., Papettia, A., Peruzzinia, M., & Germani, M. (2014). A smart home information management model for device interoperability simulation. *Procedia CIRP, 21,* 64–69.

4. Amza, C. G., Cantemir, D., Cantemir, I., et al. (2018). *Guidelines on Industry 4.0 and drone entrepreneurship for VET students.* Danmar Computer LLC ul.K. Hoffmanowej 19-35-016, Rzeszow.
5. Festo didactic cyber-physical factory, A universal Industry 4.0 training factory. A guided tour on the learning industry 4.0 system. 1058@2016/12
6. Bourey, J. P., Grangel, R., Doumeingts, G., & Berre, A. (2007). Deliverable DTG2.3 from the INTEROP project. Report on Model-Driven Interoperability. http://interop-vlab.eu. Accessed 15 May 2013.
7. Ansari, F., Hold, P., & Sihn, W. (2018). Human-centred cyber-physical production system: How does Industry 4.0 impact on decision making tasks? In *IEEE Technology and Engineering Management Conference.*
8. Kondak, K., Ollero, A., et al. Unmanned aerial systems physically interacting with the environment. Load transportation, deployment and aerial manipulation. In *Handbook of unmanned aerial vehicle* (pp. 2755–2785). Springer.
9. Michel, H. (2017). Amazon's drone patents. Center for the study of drone, September 2017.
10. Faldick, O., Payne, R., Fitzgerald, J., & Buhnova, B. (2017). Modelling systems of systems interface contract behavior. In *Formal Engineering Approaches to Software Components and Architecture (FESCA'17)*, EPTCS 245 (pp. 1–15). https://doi.org/10.4204/EPTCS.245.1.
11. Sangita, D. E., Mottok, J., & Brada, P. (2019). Towards semantic model-to-model mapping of cross-domain component interfaces for interoperability of vehicle application. In *CEUR Workshop Proceedings* (Vol. 2442).
12. Moro, G., Wanderley, P., Abel, M. -H., Paraiso, E. C., & Barthès, J. -P. (2018). MBA: a system of systems architecture model for supporting collaborative work. *Computers in Industry, 100*, 31–42.
13. Linux Foundation, The open-source project for drone Aviation Interoperability, Press release 2019
14. Zacharewicz, G., Diallo, S., Ducq, Y., Agostinho, C., et al. (2016). Model-based approaches for interoperability of next generation enterprise information systems: State of the art and future challenges. In *Information systems and e-business management.* Springer. https://doi.org/10. 1007/s10257-016-0317-8.
15. Bekke, & Liebisch, D. C. (1992). *Semantic data modelling.* Prentice Hall, Englewood Cliffs, N.J. ISBN 0-13-806050-9.
16. Bardaro, G., Sembrebon, A., Matteucci, M. (2018). A use case in model-based robot development using AADL and ROS. In 2018 *ACM/IEEE 1st International Workshop on Robotics software Engineering, RoSE'18*, May 2018.
17. Erskine, J., & Chriette, A. (2019). Control and configuration planning of an aerial cable towed system. In *International Conference on Robotics and Automation (ICRA 20019)*, May 2019.
18. Smartsoft. (2019). Smartsoft components and toolchain for robotics. University of Applied Sciences, Ulm
19. Pramsohler, T. (2015). A layer interface-adaption architecture for distributed component-based systems. *Future Generation Computer Systems, 47*, 113–126.

Internet of Things

Applying Distributed Ledger Technology to Facilitate IIoT Data Exchange: An Approach Based on IOTA Tangle

Xiaochen Zheng⑩, Shengjing Sun⑩, Joaquín Ordieres-Meré⑩, Jinzhi Lu⑩, and Dimitris Kiritsis

Abstract Data interoperability is a fundamental dimension in enterprise interoperability. The interoperability of data is concerned with exchanging information coming from heterogeneous sources among different partners. Under the Industry 4.0 context, Internet of Things (IoT) technology has been widely implemented in manufacturing factories which enables the concept of smart factory. The huge amount of Industrial IoT (IIoT) devices are generating large volume of data related to all aspects of the enterprise. The free exchange of these big IIoT data is crucial to enterprise interoperability. However, in practice, the overwhelming part of the industrial data remains siloed preventing the full use of the big IIoT data. Concerns about data security and privacy bring more obstacles to industrial data sharing. With the decentralized and consensus-driven characteristics, distributed ledger technologies (DLT), represented by blockchain, provide reliable solutions to improving enterprise interoperability. This paper explores the application of IOTA in IIoT data exchange. IOTA is a tangle-based distributed ledger designed specifically for the IoT applications. A prototype data exchange system is developed based on IOTA and its data communication protocol, masked authenticated messaging (MAM), to demonstrate the feasibility of the proposed approach.

Keywords Data interoperability · Industrial IoT · Distributed ledger technologies · Blockchain · IOTA tangle · Masked authenticated message

1 Introduction

For a modern enterprise, the ability to interoperate with partners from inside or outside is not only a quality and advantage for gaining competitiveness in the market but also becoming a question of survival [1]. According to the definition of IEEE [2],

X. Zheng (✉) · S. Sun · J. Ordieres-Meré
ETSII, Universidad Politécnica de Madrid, José Gutiérrez Abascal 2, 28006 Madrid, Spain
e-mail: xiaochen.zheng@epfl.ch

X. Zheng · J. Lu · D. Kiritsis
Institute of Mechanical Engineering, EPFL, 1015 Lausanne, Switzerland

"interoperability" means the ability for two (or more) systems or components to exchange information and to use the information that has been exchanged. IDEAS defined "enterprise interoperability" as the ability of interaction between enterprises which is achieved if the interaction can, at least, take place at three levels: data, application, and business process [3]. Enterprise interoperability makes possible of two or more enterprises (of the same organization or from different organizations and irrespective of their location) with the ability of exchanging or sharing information (wherever it is and at any time) and using functionality of one another in a distributed and heterogeneous environment [4].

The interoperations among enterprises can happen from various levels such as data interoperation, service or organization interoperation, information system (IS) or IT application interoperation, processes interoperation, and business interoperation [4, 5]. They either concern the internal business processes and services of a given enterprise or cross-organizational business processes spanning partner companies or flowing across enterprise networks. The various viewpoints of enterprise interoperations are as shown in Fig. 1 which is adapted from previous studies [4, 5].

Data interoperability is the fundamental dimension for achieving higher level and enterprise interoperability. With the wide deployment of IIoT devices, huge amount of data related to different aspects of an enterprise have been generated every moment. The efficient interoperation of these IIoT data is crucial to implement the concept of smart factory. However, in practical applications, there are many barriers preventing successful data interoperability among enterprises such as different semantics and syntax to represent information, different database technologies, and strict data management policies.

Concerns about data security/privacy issues are making data protection regulations stricter. For instance, the European Union published the General Data Protection Regulation (GDPR) [6] to protect private data which will further impede data sharing. The absence of certified authenticity and audit mechanisms during data exchange may also make data owners hesitate to share data freely. Different type of attacks, such as

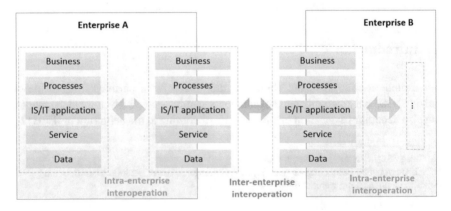

Fig. 1 Enterprise interoperation levels (based on previous studies [4, 5])

"man-in-the-middle" attacks and data tampering, can also occur when sharing data using traditional protocols and databases [7]. For industrial application scenarios, the IIoT data are usually with high frequency and require real-time exchange. Therefore, the exchange process needs to be very cheap or even totally free, which is difficult to realize using traditional technologies [8].

To address above-mentioned concerns, innovative methods are needed to establish data access policies among arbitrary parties, accommodating new participants and request types dynamically [9]. The repaid development of distributed ledger technologies (DLT) provides a possible solution to this challenge. A distributed ledger is a distributed database, maintained by a consensus protocol run by nodes in a peer-to-peer network without any central administrator [10]. Blockchain has been one of the most popular DLT in recent years due to the success of cryptocurrencies in financial field like Bitcoin [11]. Blockchain technology has been applied to variety of domains and gained mainstream attention due to some unique features, such as decentralized control, high anonymity, and distributed consensus mechanisms [12–14]. The adoption of blockchain in a data sharing system could enable better data control and makes possible of fine-grained tracking of different data usages [15].

Although many studies and projects have proved the practical value of blockchain technologies like cryptographic currencies [11] and smart contracts [16], these protocols still have various limitations that make them inadequate for IIoT data sharing.

- **Scalability** Transaction rate, i.e., the number of transactions processed per second over the whole network of a blockchain has an inherent limit, because all transactions must be attached to the longest chain causing the "blockchain bottleneck" issue [17]. For example, it has to wait up to six blocks for a transaction to be approved before reaching a high level of confidence on the Bitcoin network [15, 18]. The transaction rate of Bitcoin protocol has been lower than six transactions per second in the whole network during most of the time in the year 2019 [19]. Similarly, the Ethereum protocol processed about ten transactions per second across the entire network even after the upgrade in 2019 [20]. This low transaction rate is far away the requirements of industrial machine-to-machine data exchange scenarios.
- **Transaction Fees** The transaction fees, no matter the value of the transaction itself, is another main drawback when applying blockchain in industrial scenarios. For example, the Bitcoin protocol requires a fee that may exceed \$0.30 for each transaction according to the latest statistics [21]. Currently, it is impossible to remove these fees in the blockchain platform as they provide motivations for the creators of blocks [22]. These high transaction fees make no sense for the high-frequency data exchange in IIoT environment. It is highly possible that the transaction fee is higher than the value being transferred which makes no sense.
- **Centralization** Blockchain is designed to be decentralized, but a lot of computing power is required to create blocks. In practical, large part of the mining power has been controlled by some mining pools making blockchain centralized to some extent. The latest statistic shows that the seven largest mining pools control 77.1%

of the of the network's mining power (F2Pool 17.3%, Poolin 15.1% BTC.com 13.8%, AntPool 9.7%, ViaBTC 7.4%, BTC.TOP 6.3%, SlushPool 4.3%, BitFury 3.2%) [23].

- **Vulnerable to Quantum attack** Quantum computing, although still a hypothetical construct currently, has been proved feasible. A quantum computer is supposed to be supper efficient for solving problems that depend on trial and error to find a solution [22]. Blockchains that are based on proof-of-work, such as Bitcoin, are vulnerable to quantum computing attacks. Theoretically, a quantum computer could be billions of times more efficient than a classical computer when mining the Bitcoin blocks [24], which would enable it to control over 51% of computing power of the whole network and possible to breakdown the entire network.

In order to take the advantages of Blockchain technology and meanwhile overcome the above-mentioned limitations, a new DLT is needed. In this paper, we adopt a tangle-based DLT paradigm which is designed specifically for the IoT industry, named IOTA, to facilitate the IIoT data exchange among different stakeholders.

The rest of this paper is organized as follows. Section 2 introduces the methodology we used. An application framework is developed, and some main enabling technologies are explained. A prototype system is demonstrated, and an exemplary experiment is conducted in Sect. 3. The conclusion of this paper and the future work are introduced in Sect. 4.

2 Methodology

The efficient data interoperation among different IoT environments require frequent and automatic machine-to-machine data exchange system. Conventional blockchain protocols like Bitcoin blockchain and Ethereum smart contract cannot fulfill the requirements of IIoT data interoperation scenarios due to the limitations mentioned in previous sections.

In this paper, we utilized a tangle-based DLT protocol which is specifically designed for the industrial data exchange scenarios, named IOTA. It succeeds the advantages of blockchain and at the meantime overcomes some of its fundamental limitations [25]. The tangle uses a directed acyclic graph (DAG) for storing transactions instead of sequential blocks. To issue a new transaction in the tangle, users must perform a small amount of computational work to approve two previous transactions, and this new transaction will be validated by some subsequent transactions [22]. This structure allows high scalability as more transactions joined in the tangle, the faster transactions can be approved. Moreover, financial rewards can be eliminated owing to the unique validation method enabling completely fee-free transactions with IOTA. This is extremely important for IIoT data interoperability. Furthermore, no miners exist in IOTA tangle; therefore, it is truly decentralized.

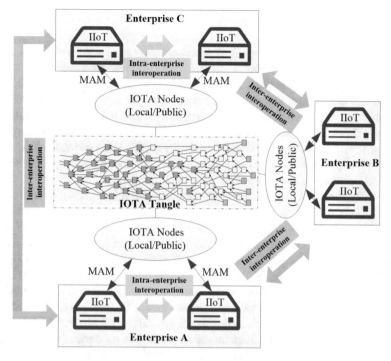

Fig. 2 Intra-enterprise and inter-enterprise IIoT data exchange framework based on the IOTA tangle and MAM protocol

As shown in Fig. 2, a framework for intra-enterprise and inter-enterprise IIoT data interoperation is proposed based on the IOTA tangle and some relevant enabling technologies. In this framework, IIoT data are interchanged between data publishers and data subscribers. Any device with basic computing capability and access to the internet, like a computer, a smartphone, single-board computer, or any IoT device connected to a gateway, can be a data publisher or subscriber. A data publisher or subscriber can be from inside an enterprise for intra-enterprise interoperation, or from another enterprise for inter-enterprise interoperation. A data publisher can publish different kind of data to the tangle using different encryption and privacy protocols, which will be explained in the following section. The data are published in their own channels, and each channel has an address. The subscribers of a data channel will receive the new published data. The published data could be encrypted in which case an extra decryption key will be required to decrypt the received message. In the tangle, the data publishing and receiving are processed through IOTA nodes. A node can be a computer or server connected to the IOTA network. Users could configure their own local nodes or use public nodes. A user can be a data publisher or a subscriber or both at the same time.

Another key-enabling tool for the proposed framework is a data communication protocol of IOTA, named masked authenticated messaging (MAM). It supports publishing and receiving an encrypted data stream over the tangle regardless of the size or cost of device [26]. MAM uses channels for data distribution which works similar to the radio broadcasting. Data publishers can create a channel with a unique address and publish a message at any time. In order to spread the message through the network and prevent spamming, the node publishing the message needs to conduct a small amount of proof-of-work. The users who are interested in this message can subscribe this channel with the address and receive the message. Merkle hash tree (MHT) is adopted as the signature scheme in the MAM protocol to encrypt the message [26]. The address of a channel is the root of this MHT which itself is created using the unique identification of the user. More details about the signature scheme have been introduced in our previous work [27].

MAM supports three privacy and encryption modes, i.e., public, restricted, and private, to control the access to a channel [26]. In public mode, the root of the MHT is used directly as the address of the MAM channel and the key to decode the message. Any user who knows the address, even randomly, will be able to decode and consume the message. In private mode, the hash of the MHT root is used as the address of the channel, while the message is encrypted using the root. In this mode, only the publisher can decode and consume the message. In restricted mode, an authorization key is added based on private mode. The address of the channel is the hash of the key and the root. In this mode, subscribers of the channel can receive the encrypted message with the channel address and decrypt it with the authorization key. The restricted mode is the most commonly used for IIoT data exchange because it enables a message publisher to revoke access to future messages from subscribers by changing the authorization key without changing the channel address.

3   Prototype and Experiment

In order to verify the feasibility of the proposed framework and explore the implementation process in reality, a prototype system has been developed. The MAM-enabled data publishing and receiving functions were realized based on the JavaScript library provided by the IOTA foundation (https://github.com/iotaledger/mam.client.js) which is open available. More technical details about publishing and receiving data over the tangle using MAM are introduced in our previous work [27], and the complete JavaScript codes are available online and ready to be reused (https://github.com/zhengxiaochen/iota_mam_data_sharing).

A series of experiments have been conducted using the prototype system to publish and receive sensor data collected from different IIoT devices. For example, one of the experiments focused on the environmental quality data interoperation within a steel manufacturing factory. Figures 3 and 4 show two examples of published messages using public and restricted MAM mode, respectively. In public mode, the address of the message (second line) is the same as the MHT root (first line), as shown in Fig. 3; while in restricted mode, the message address is the hash of the MHT root, which is different from the address, as shown in Fig. 4. Data consumers must know both the

```
Root:   KTNURJVDJTPBIVFBFMVLVQTZBEUULUUBLSXXVBNBYSOGAHEPTFVKFJRCSLC9CTCQHJVZUCJGNZOVXYPUU
Address: KTNURJVDJTPBIVFBFMVLVQTZBEUULUUBLSXXVBNBYSOGAHEPTFVKFJRCSLC9CTCQHJVZUCJGNZOVXYPUU
mam_mode:public
waiting_time:25852
location: Celsa Group Office, timestamp: 2019-01-09 00:00:00, pm2_5: 10.5, pm10: 10.833,tvoc: 0
.034, co2: 0.2, temperature: 26.7, humidity: 14.925,illumination: 0.0, noise: 75.257, hcho: 0.0
2, co: 0, c6h6: 0.0, no2: 0, o3: 0
```

Fig. 3 Sensor data published to the tangle with public MAM mode

```
Root:   DTHVYM9BMUHA9AKJEVFPNDPFNZVCQTIN9IQLPXYIALKZMAXAEBIOVKVTZWVBRNBDKJND9QWRDEWHJGTXE
Address: CGMRUNTJRHWJCQRKFZUIUGXTNMBESVECSKMHXWDNCFTCOYHQLZIJDQEJCYDYV99QCGSIKOPY9WETEFKYN
mam_mode:restricted
waiting_time:16062
location: Celsa Group Office, timestamp: 2019-01-09 00:00:00, pm2_5: 10.5, pm10: 10.833,tvoc: 0
.034, co2: 0.2, temperature: 26.7, humidity: 14.925,illumination: 0.0, noise: 75.257, hcho: 0.0
2, co: 0, c6h6: 0.0, no2: 0, o3: 0
```

Fig. 4 Sensor data published to the tangle with restricted MAM mode

address to receive the message and the authorization key to decrypt it in restricted mode. If the publisher wants to withdraw the authorization in the future, it can change the authorization key at any time to the new published messages; by doing this, the subscribers without the new authorization key will not be able to decrypt the new message which means they will also lose the access to the future ones. This feature provides the data publisher with granular control over the shared data, which could bring great benefits to the IIoT data exchange and make IOTA distributed ledger outperform traditional block-based ledgers.

4 Discussion

This paper proposed a novel IIoT data interoperation framework utilizing the emerging distributed ledger technology. After analyzing the advantages and limitations of traditional block-based ledgers, we introduced the tangle-based IOTA distributed ledger to address the concerns of enterprise about data security/privacy and the lack of ensured authenticity/audit trails. Designed specifically for the IoT industry, IOTA could provide a scalable, lightweight, and zero-fee secure communication and transaction protocol for IIoT interoperation. A prototype system was developed under the proposed framework to demonstrate the implementation process in practice and to verify the feasibility of the proposed method. Experiment results showed that the proposed system could provide granular access controls to different sensor data by combining public and restricted MAM protocols. The proposed approach could greatly facilitate both intra-enterprise and inter-enterprise data interoperability. It also provides solutions to help handle the big IIoT data generated by numerous IoT devices in Industry 4.0 era and makes possible of the promising smart manufacturing.

Although the current implementation of IOTA is already usable, some limitations still exist. One of the main drawbacks is the presence of the coordinator in the current network, which will be removed in the future. It is introduced temporarily to secure the tangle network by issuing milestone transactions which will refer and approve all trustworthy tractions in the network. However, the existence of coordinator makes the IOTA not fully decentralized as designed and may cause a single-point failure. Another disadvantage of IOTA is that it does not support decentralized applications as the Ethereum smart contracts do. Although this drawback has no major impact on the data exchange application in this study, it limits the wide application of IOTA in other domains.

Currently, IOTA tangle and its MAM protocol are under development and are evolving rapidly. As more nodes are connected to the tangle network and continuous development efforts spent, some of the afore-mentioned limitations of IOTA will be solved and the performance of IOTA tangle is expected to improve greatly soon.

Acknowledgements The authors thank the financial support by the EU Commission within the research projects QU4LITY Digital Reality in Zero Defect Manufacturing (EU H2020 825030) and FACTLOG-Energy-aware Factory Analytics for Process Industries (EU H2020 869951).

References

1. Chen, D., & Daclin, N. (2006). Framework for enterprise interoperability. In *Proceedings of IFAC workshop EI2N: Bordeaux*.
2. Geraci, A., Katki, F., McMonegal, L., et al. (1991). *IEEE standard computer dictionary: Compilation of IEEE standard computer glossaries*. IEEE Press.
3. IDEAS Consortium. (2003). *IDEAS project deliverables (WP1-WP7)*. Public Reports.
4. Vernadat, F. B. (2010). Technical semantic and organizational issues of enterprise interoperability and networking. *Annual Reviews in Control, 34*(1), 139–144.
5. Guglielmina, C., & Berre, A. (2005). *ATHENA, "Project A4"(slide presentation)*. ATHENA Intermediate Audit.
6. Regulation GDPR. (2016). Regulation (EU) 2016/679 of the European Parliament and of the Council of 27 April 2016 on the protection of natural persons with regard to the processing of personal data and on the free movement of such data, and repealing Directive 95/46. *Official Journal of the European Union (OJ), 59*(1–88), 294.
7. Callegati, F., Cerroni, W., & Ramilli, M. (2009). Man-in-the-middle attack to the HTTPS protocol. *IEEE Security & Privacy, 7*(1), 78–81.
8. Sønstebø, D. *IOTA data marketplace*. https://blog.iota.org/iota-data-marketplace-cb6be4 63ac7f. Last accessed 2019/11/05.
9. Ordieres-Meré, J., Villalba-Díez, J., & Zheng, X. (2019). Challenges and opportunities for publishing IIoT data in manufacturing as a service business. In *25th international conference on production research manufacturing innovation*. Cyber Physical Manufacturing.
10. Brogan, J., Baskaran, I., & Ramachandran, N. (2018). Authenticating health activity data using distributed ledger technologies. *Computational and Structural Biotechnology Journal, 16*, 257–266.
11. Nakamoto, S. (2008). Bitcoin: A peer-to-peer electronic cash system.
12. Böhme, R., Christin, N., Edelman, B., et al. (2015). Bitcoin: Economics, technology, and governance. *Journal of Economic Perspectives, 29*(2), 213–238.
13. Ali, S. T. (2015). Bitcoin: Perils of an unregulated global P2P currency (transcript of discussion). In *Cambridge international workshop on security protocols*. Springer.

14. Harlev, M. A., Sun Yin, H., & Langenheldt, K. C., et al. (2018). Breaking bad: De-anonymising entity types on the bitcoin blockchain using supervised machine learning. In: *Proceedings of the 51st Hawaii international conference on system sciences.*
15. Mamoshina, P., Ojomoko, L., Yanovich, Y., et al. (2018). Converging blockchain and next-generation artificial intelligence technologies to decentralize and accelerate biomedical research and healthcare. *Oncotarget, 9*(5), 5665.
16. Ethereum Foundation. *Decentralized autonomous organization.* https://www.ethereum. org/dao. Last accessed 2019/11/05.
17. IOTA Foundation. *Meet the Tangle.* https://www.iota.org/research/meet-the-tangle. Last accessed 2019/11/05.
18. Bitcoin Wiki. *Confirmation.* https://en.bitcoin.it/wiki/Confirmation. Last accessed 2019/11/05.
19. Transaction Rate. https://www.blockchain.com/en/charts/transactions-per-second. Last accessed 2019/11/05.
20. Ethereum transaction chart. https://etherscan.io/chart/tx. Last accessed 2019/11/05.
21. Bitcoin Avg. *Transaction fee historical chart.* https://bitinfocharts.com/comparison/bitcoin-tra nsactionfees.html#1y. Last accessed 2019/11/05.
22. Popov, S. (2018). The tangle.
23. Pool Distribution. https://btc.com/stats/pool?pool_mode=year. Last accessed 2019/11/05.
24. Brassard, G., Høyer, P., & Tapp, A. (1998). *Quantum cryptanalysis of hash and claw-free functions.* Springer.
25. IOTA Foundation. *What is IOTA?* https://www.iota.org/get-started/what-is-iota. Last accessed 2019/11/05.
26. Handy, P. *Introducing masked authenticated messaging.* https://blog.iota.org/introducing-mas ked-authenticated-messaging-e55c1822d50e. Last accessed 2019/11/05.
27. Zheng, X., Sun, S., Mukkamala, R. R., et al. (2019). Accelerating health data sharing: A solution based on the Internet of Things and distributed ledger technologies. *Journal of Medical Internet Research, 21*(6), e13583.

Analysis of Data Exchange Among Heterogeneous IoT Systems

Jannik Laval, Nawel Amokrane, Mustapha Derras, and Néjib Moalla

Abstract Data interoperability allows data exchanges among information systems, their sub-systems and their environment. The multiplicity of these exchanges and the increasing amount of exchanged data can generate dysfunctions with negative impact on the overall performance of the communicating systems. Data interoperability should therefore be continuously assessed and improved. We propose a messaging metamodel that aggregates collected information from several pub/sub-communication protocols, and we present a work in progress which utilizes services provided by AMQP, MQTT, CoAP and Kafka to collect information in order to analyze data exchanges. Including these pub/sub-communication protocols and the data analysis platform Moose to achieve monitoring, we propose the pulse framework that provides a tracking of architecture changes in the pub/sub-systems. We analyzed the differences between the protocols to provide a generic metamodel to include all of these pieces of information in the same system. It will allow to extract precise information about the evolution of the system.

Keywords Data interoperability · Data analysis · Monitoring · Message brokers

1 Introduction

The problem of interoperability between heterogenous systems already exists and is amplified by the strong deployment of Internet of Things. To respond to this challenge, enterprises address this problem by emphasizing on the use of open standards for data format as well as communication protocols. Despite these efforts, interoperability is still a real issue that cannot be ignored.

J. Laval (✉) · N. Moalla
University Lumière Lyon 2, DISP lab EA4570, Bron, France
e-mail: Jannik.Laval@univ-lyon2.fr

N. Amokrane · M. Derras
Berger-Levrault, Lyon, France
e-mail: Nawel.amokrane@berger-levrault.com

B. Archimède et al. (eds.), *Enterprise Interoperability IX*, Proceedings of the I-ESA
Conferences 10, https://doi.org/10.1007/978-3-030-90387-9_7

73

Distributed message brokers are typically used to decouple separate stages of a software architecture. They permit communication between these stages asynchronously, by using the publish/subscribe (pub/sub) paradigm. Implementing a message-oriented middleware enables asynchronous communication which allows applications to be more loosely coupled. As a result, available resources can be better utilized and systems performance improved. These message brokers are also finding new applications in the domain of IoT devices and may also be used as a method to implement an event-driven processing architecture.

The increase in the number of exchanges and consequently in the amount of data leads to the need to deploy monitoring and analysis systems. During exchanges in an event-oriented system, monitoring can be carried out at different levels (e.g., at the level of a node, an exchange, messages, etc.) or even on the entire system. Several different tools are needed. For example, RabbitMQ offers a management console to monitor the status of messages and the status of each element of an AMQP system. This console shows the list of resources, their characteristics, and some statistics. This console is suitable when the maintainer focuses his analysis on a particular node and knows the structure of the system. For example, when a queue is accessed, the messages in that queue are displayed. This allows an analysis of the situation at a particular time. However, when you want to do more complex analysis, advanced queries on resources, then these tools are not adapted. They require time-consuming manual work. For example, consumed messages are no longer presented, so it is not possible to follow their evolution. The entities in the structure (exchanges, waiting lines, etc.) also disappear from the management console as soon as they are deleted. Thus, when a consumer disconnects, these elements also disappear and are no longer usable. The management console does not allow you to view the history of a system's resources.

When considering existing open-source and monitoring tools (Nagios, Zabbix) that are great enterprise level software designed to monitor everything from performance, availability of servers, network equipment to web applications and database, we notice that they are capable of monitoring components like network protocols, operating systems, web server, website, middleware and so on, but only focusing on low level monitoring information such as, performance indicators or memory usage. Our approach uses monitoring for the assessment of interoperability, an analysis capable of defining a classification of potential causes by order of importance for a given problem. A monitoring system is defined as a process or a set of distributed processes including collection, interpretation and dynamic processing of information related to an application being run.

The messaging data model presented in Amokrane et al. [1] aggregates data collected related to message exchanges and is created for the Berger Levrault messaging infrastructure. It provides a common control point and facilitates the extraction of interoperability related indicators. The messaging data model describes the messaging structure implemented through message queueing and exchange system. It is used to collect meta-information from log services offered by the exchange infrastructures and keeps track of the exchanged messages.

In this paper, we extend the messaging metamodel to consider a more generic model adapted to messaging paradigms. We consolidate it by analyzing AMQP broker, MQTT broker, KAFKA broker and CoAP server. We also propose the pulse framework that uses this model to collect meta-information to be able to (i) keep track of the exchanged messages, (ii) simplify the visualization of exchanges, (iii) enhance the maintainability by detecting exceptions (ex: problem of transfer of a message), precising of the context and the origin of the problem and providing alerts and notifications. The pulse framework integrates dynamic features, where the lifecycle of different components of the architecture is depicted by including creation and deletion dates as well as timestamps.

In the remaining sections, Sect. 2 presents related work. Section 3 exposes the pulse framework and related tools that enable the evaluation of data interoperability. Section 4 describes its underling metamodel. Implementation is described in Sect. 5. Section 6 concludes this article and opens perspectives.

2   Related Work

Interoperability assessment evaluates the ability of enterprises or systems to undertake common activities or exchange data. Several interoperability assessments approaches have been proposed since the emergence of the concept of interoperability: maturity models (LISI, LCIM, OIM), interoperability score [2] or degree of interoperability [3]. However, these methods do not allow to precisely indicate or locate interoperability problems and mainly focus on general notions. Also, few interoperability assessment methods address the effective (post implementation) evaluation of data interoperability and few are tooled [4]. These methods have nonetheless provided the fundamental concepts that allow formalizing and evaluating interoperability by indicating whether interoperability problems exist or not. Based on these concepts, other approaches [5, 6] have defined a set of interoperability requirements (e.g., "partners provide permissions for data updates," "received data is conform to required data") that should be verified to achieve interoperability.

In terms of existing tools allowing monitoring, we can mention: ELK Stack [7] and Qlik Sense. ELK Stack is the combination of three open-source tools Elasticsearch, Logstash and Kibana. Elasticsearch is a No-SQL database with a focus on search and analysis capabilities, Logstash is a log aggregator that gathers data from different sources, transforms, enhances it and sends it to different output destinations and Kiabana is a visualization tool that works on top of Elasticsearch. Qlik Sense is a dashboard solution that enables one to visualize and preform data analytics. It supports interactive and dynamic visualizations; it is flexible and multiplatform.

3   The Pulse Framework

To explore communication rules allowing to operate connected objects and to reach
data exchanges monitoring, visualization and adaptation through pub/sub-messaging
model AMQP, MQTT, KAFKA and CoAP protocols, the general prototype framework is demonstrated in Fig. 1. It represents the collected data from different protocols
related to the exchange of messages and the log of events. The proposed monitoring
system is composed of four layers: (i) data importing and model population layer,
(ii) time management and model versioning layer, (iii) persistence layer and (iv)
visualization and analysis layer.

3.1   Data Importing and Model Population Layer

One of the main challenges is to import data from different sources with different
formats. For that, we define a metamodel representing the structure of a distributed
exchange system. The metamodel is detailed in the next section. The dedicated
importers take data as input and instantiate the model with the information.

3.2   Time Management and Model Versioning Layer

The instantiation of the messaging model (via the different collecting components)
revealed the issue of the historization of the versions of the model to take into account

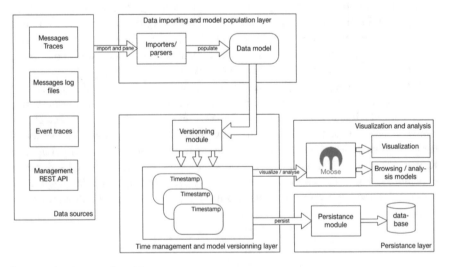

Fig. 1 Pulse framework architecture

the dynamic aspects of the system. We need a kind of historization to be able to understand the events in earlier versions of the architecture and to favor a better analysis for maintenance needs.

A trivial method would be to integrate a timestamp for creation, deletion and update to each entity of the model. The problem with this approach is the strategy to build a specific status at a specific time. Another method can be the creation of a new model each time there is an update, which takes big space at each new data coming from the importers.

We consider another solution based on Orion [9]. Orion is a model that enables creating different versions of a data model considering the tracking of changes in this model. The principle behind Orion is that each change triggers an Orion action which is responsible for adding the change to the data model. Each change can result in an updated version of the data model. Orion optimizes the persistence of different versions of the model, where Orion handles deltas and pointers to earlier versions. Orion copies the sole entities that have been impacted by a change. Figure 2 illustrates our versioning strategy. This version management of the data model allows us to follow the evolution of the messaging architecture over time, where each version represents an image of the architecture at a given moment.

To resume, an Orion version includes the latest changes and information about the action that was preformed to create this version. For the user, each version represents

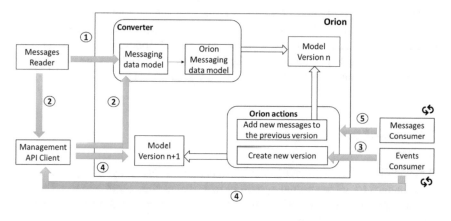

Fig. 2 Orion and the model versioning process

a screenshot of the monitored system in a specific time. In other words, instead of having an overcrowded unusable model, Orion provides multiple small models, each of them describing a change to the system at a certain time.

We defined a strategy to create a version each time it is necessary. In the case of a message exchange system, we define two kinds of events.

- A change in the architecture or to the configurations/settings of the monitored system (queue creation, queue deletion, user permissions changed, etc.). The status of the system before and after the change must be kept. So, for each of these changes, an Orion version is created.
- A new trace (new message published, message received, new connections, etc.). In this case, the framework instantiates a dedicated entity in the current version of the model. It is not necessary to create an updated version.

3.3 Persistence Layer

Orion keeps different versions of the model in the Pharo[1] image of Moose, and due to the way that Orion versioning system works and the fact that entities that do not change are not copied from version to version but rather a reference to unchanged entities in the previous version is copied, storing different versions in memory will not pose a space problem, but for the long run, we needed a way to store our models in a persistent way. With this feature, we enable external systems like Grafana[2] to acquire metrics and use them to display certain visualizations.

When our persistence module is called, it stores all versions of the Orion model to a json file. It can be extended to output other kind of structured data.

3.4 Visualization and Analysis Layer

Implementing model versioning enabled us to preform two things: analyze and visualize changes to the monitored system in real time as they occur and to go back in time to a previous state of the monitored system to visualize and analyze changes and their impact at a giving time.

These two features allow us not only to detect interoperability issues as they occur but also to identify the potential source of a certain problem by going back to previous states.

---

[1] https://pharo.org/.

[2] https://grafana.com/.

4 Pulse Metamodel

The goal of the pulse metamodel (Fig. 3) is to represent three aspects of the messaging structure:

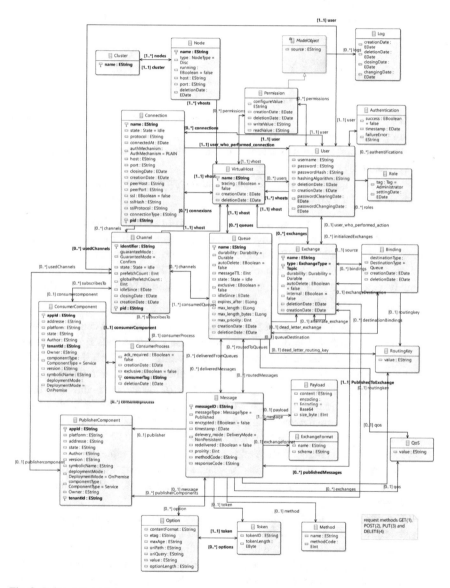

Fig. 3 Pulse metamodel

Table 1 Comparing concepts with the AMQP-based model

Protocols	Similar concepts	Different concepts
MQTT	Cluster, V-host, user, connection, channel, exchange, binding, queue, routing keys, message, security	QoS, persistent session
Kafka	Cluster, user, connection, channel, message, security	Topic, partition, QoS
CoAP	Cluster, V-host, user, connection, channel, exchange, binding, queue, routing keys, message, security	QoS, token, options, request methods

- A static representation: the messaging structure implemented through message queuing and exchange system.
- A dynamic representation, where the messages flow from publishers to consumers is represented.
- The lifecycle representation of architecture, where components (e.g., queue, exchange, …) are created, modified, deleted.

This metamodel is aimed to be generic enough to consider different protocols, as these protocols can be used to set up data exchanges within information systems and with their environment. We extend a previously presented metamodel [1] mainly based on AMQP with other protocols: AMQP, MQTT, Kafka and CoAP protocols. The analysis of these protocols allowed as to determine similarities and differences comparing to AMQP (Table 1).

- AMQP is a standard protocol for exchanging messages between applications. It is message-oriented and allows asynchronous communication using queues. It makes applications interoperable by facilitating communication and offers message encryption. The protocol is used in client/server communications and in the management of IoT devices. The structure of the system is message-based: The message is created by a message producer, passes through an exchange and transits through queues according to routing keys to finally reach the message consumer. Each message contains a header and a content. The header is formatted according to an exchange format and contains a set of metadata useful for routing the message. The content contains the message to be processed by the message consumer. The message broker manages all these elements. Clients connect using identifiers.
- MQTT [9, 10] is a protocol that enables specialized message exchanges for lightweight machine to machine and IoT communications. Like AMQP, it is a publish/subscribe protocol. It works in a similar way to AMQP. Some differences are: It has a payload containing the message, and a minimalist header. A session indicator shows whether a persistent session has previously been created with the client. The real difference with AMQP is the presence of a quality-of-service parameter. The broker allows one to choose one of three levels of service. This level of service ensures that a message has reached its destination.

- Kafka [11, 12] is a communication protocol optimized for rapidly distributing messages between applications in a scalable manner. It is a publish/subscribe protocol with a message storage system designed as a distributed transaction log. This system allows for continuous data processing. Thus, a record in the transaction log contains a key, a timestamp and the content of the message. A Kafka cluster stores these transactions in categories called topics, using the same principle as AMQP. Topics are organized into partitions. Each broker can have several partitions. A partition is an ordered sequence of records.
- CoAP [13] is a communication protocol specialized in constrained systems. It has been defined in RFC 7252. CoAP's main objective is to address the constraints of specific environments, e.g., energy management in building automation. CoAP is based on the UDP protocol with an overlay for sending messages and managing responses. A message can be of four different types: confirmable, unconfirmable, acknowledgment and reset. Thus, in the header of CoAP messages, a message code or a response code is included.

Based on the analysis of these four protocols, we propose a metamodel that can be instantiated independently on each of them. The metamodel is presented in Fig. 3. In this figure, each item is represented. To include topic and partition items of Kafka, we choose to use the same concepts as the other protocols: Queue is used for partition concept. Exchange is used for topic concept.

5 Implementation

The pulse metamodel is implemented in the Moose framework [4]. This framework is a data analysis platform. It contains data import tools, modeling tools, a domain specific language allowing queries to be made on these data and a language for creating visualizations.

The implementation of pulse contains all the elements presented in the framework previously shown in Fig. 1. Four importers retrieving data from RabbitsMQ have been developed: two message importers and two others allowing the model to be synchronized with the architecture of the RabbitMQ application.

Thus, the tool is able to collect:

- Message traces from the RabbitMQ log files. These traces are read thanks to two modules: a message trace interpreter intervening on the log file a posteriori and a consumer subscribed to a message trace exchange. The latter is more reactive, but intrusive because it adds a listening node in the system.
- The history of events in the life cycle of the system's components, from the RabbitMQ Event Exchange plug-in. This information is collected through the instantiation of a consumer subscribed to a dedicated event exchange.
- The configuration of RabbitMQ, from a REST client that queries the RabbitMQ management API.

6　Conclusion

This article presents a framework for analyzing the exchanges of a distributed information system. This framework is composed of importers, allowing the collection of data from EDA systems such as AMQP, CoAP, MQTT and Kafka. It uses a generic metamodel implemented in Moose and offers analysis tools to identify information exchanges in the system. It allows to analyze these exchanges to identify potential problems or to visualize the active and inactive nodes of an IoT system. The model takes into account the dynamic aspects of the system by considering the messages and by modeling the changes in the architecture over time.

The next step is to consider the whole tooling: useful visualizations, queries and importers. Then, we work on the detection of system failures from the analysis of messages and the treatment of these failures. Another issue is to process the mass of data and to use known big data strategies for this.

References

1. Amokrane, N., Laval, J., Lanco, P., Derras, M., & Moala, N. (2018). Analysis of data exchanges, contribution to data interoperability assessment. In *International Conference on Intelligent Systems* (IS), Madeira Island, Portugal.
2. Ford, T., Colombi, J., Graham, S., & Jacques, D. (2007). The interoperability score. Technical Report, Air Force Inst of Tech Wright-Patterson AFB OH.
3. Daclin, N., Chen, D., & Vallespir, B. (2008). Methodology for enterprise interoperability. *IFAC Proceedings Volumes, 41*(2), 12873–12878.
4. da Silva, G., Leal, S., Guédria, W., & Panetto, H. (2019). Interoperability assessment: A systematic literature review. *Computers in Industry, 106*, 111–132.
5. da Silva Serapiao Leal, G. (2019). Support à la décision pour l'analyse de l'interopérabilité des systèmes dans un contexte d'entreprises en réseau. PhD thesis, Université de Lorraine
6. Mallek, S., Daclin, N., & Chapurlat, V. (2012). The application of interoperability requirement specification and verification to collaborative processes in industry. *Computers in Industry, 63*(7), 643–658.
7. Berman, D. (2019) The complete guide to the ELK Stack. Logz.io. Available at: https://logz.io/learn/complete-guide-elk-stack/. [Consulted on 07, June 2019].
8. Laval, J., Denier, S., Ducasse, S. &, Falleri, J.-R. (2011) Supporting simultaneous versions for software evolution assessment. *Science of Computer Programming, 76*(12), 1177–1193. ISSN 0167–6423, https://doi.org/10.1016/j.scico.2010.11.014.
9. MQTT—OASIS standard. [Online]. Available: http://mqtt.org/.
10. Message, O., Telemetry, Q., & Mqtt, T. (2014). "MQTT Version 3.1.1," no. October, 2014.
11. Freiknecht, J., Papp, S., Freiknecht, J., & Papp, S. (2018). Apache Kafka. In *Big Data in der Praxis*.
12. "Apache Kafka." [Online]. Available: https://kafka.apache.org/.
13. Shelby, Z., Hartke, K., & Bormann, C. (2014). The constrained application protocol (CoAP). *Internet Eng. Task Force*, 1–112
14. Hong, X. J., Sik Yang, H., & Kim, Y. H. (2018). Performance analysis of RESTful API and RabbitMQ for Microservice web application. In *9th International Conference on Information and Communication Technology Convergence: ICT Convergence Powered by Smart Intelligence, ICTC 2018*, 2018, pp. 257–259.

Implementing Semantic Interoperability in Cloud Collaborative Manufacturing: A Demonstration Case for an Ontology-Based Asset Efficiency Testbed

Jaime Pereira⬡, Daniel Pimenta, Daniel Dias⬡, Paula Monteiro⬡, Francisco Morais⬡, Nuno Santos⬡, João Pedro Mendonça⬡, Fernando Pereira, and João P. Carvalhal

Abstract Industry 4.0 provides intelligent factories, intelligent processes, and cyber-physical systems. Systems of the future will have to be able to handle adversities autonomously. Nowadays, engineering practices are increasingly distributed and decentralized, thus causing challenges to the level of interoperability between the various systems developed. Regardless of the structure of the databases, it is necessary to have a mechanism that guarantees the interoperability between these systems. In this paper, we present two types of integrations through ontologies: vertical integration that is a way to achieve semantic interoperability between industrial plant, MES, and ERP and horizontal integration to achieve interoperability throughout the product lifecycle. Finally, this interoperability contribution was crucial to develop an asset efficiency system.

Keywords Semantic interoperability · Cloud manufacturing · Ontologies · Asset efficiency

1 Introduction

One of the greatest challenges faced nowadays is how to deal with great volumes of data coming from an increasing number of different sources. The capture of information is easier, but knowing how to do it is far harder. Newly developed architectures have focused in higher availability and affordability of sensors, in ways of acquiring data and computer networks [1]. Consequently, it has the number of uses

J. Pereira · D. Pimenta · D. Dias · P. Monteiro · F. Morais · N. Santos (✉) · J. P. Mendonça
CCG/ZGDV Institute, Guimarães, Portugal
e-mail: nuno.a.santos@algoritmi.uminho.pt

F. Pereira
Inocam, Soluções de Manufactura Assistida por Computador, Lda., S. J. da Madeira, Portugal

J. P. Carvalhal
Vanguarda—Soluções de Gestão e Organização Empresarial, Lda., M. da Maia, Portugal

© The Author(s), under exclusive license to Springer Nature Switzerland AG 2023
B. Archimède et al. (eds.), *Enterprise Interoperability IX*, Proceedings of the I-ESA Conferences 10, https://doi.org/10.1007/978-3-030-90387-9_8

sensors, networked machines are fast-growing, and in parallel, higher volumes data are generated, i.e., big data [2].

Industry 4.0 (I4.0) or Industrial Internet of Things (IIoT) aims to promote an increase in industrial productivity and efficiency through an integration of different systems, which leads to a need for integrating different software systems either at business or at manufacturing levels, inside a single plant or within a networked enterprise. Cloud computing has provided infrastructure for centralizing this information.

Different enterprise systems need to share information between each other. However, it is many times the case that data is stored, processed, and communicated in different ways by several and heterogeneity systems. Problems of misunderstanding and loss of semantic information may arise when exchanging information between them. This phenomenon is the so-called babel tower effect [3]. This effect induced by the heterogeneity of distributed systems and different domains may lead to loss of information. This is an interoperability problem.

IEEE defines interoperability as "*the ability of two or more systems or components to exchange information and to use the information that has been exchanged*" [4]. To avoid loss of information, it is required to address semantic interoperability between legacy components with different data models.

Ontologies are a way to solve interoperability problems. An ontology is a representation vocabulary, often specialized to some domain or subject matter. More precisely, it is not the vocabulary as such that qualifies as an ontology, but rather the terms in the vocabulary intended to be captured. The term ontology is sometimes referred as the body of knowledge describing a domain [5]. Ontology modeling, namely in IIoT/I4.0 settings, may refer to different integrations for supporting digital twin and digital thread lifecycles [6], namely product (vertical) and application (horizontal integration).

Thus, this paper proposes an approach for defining an ontology for both vertical and horizontal integrations able for supporting digital thread concept. ISO10303 (or Standard for Exchange of Product Data—STEP) [7] allows supporting horizontal integration. ISA-95 [8] allows supporting vertical integration. It is also needed to ensure tolling support for making use of the ontology, where typically ontology-based database access (ODBA) [9] is used. Orchestration of the data flows for a collaborative manufacturing is afterward enabled by a cloud computing architecture. This research was conducted under the project "*PRODUTECH-SIF—Soluções para a Indústria do Futuro*" (Solutions for the Industry of the Future), which is used as a demonstration case.

This paper is structured as follows: Sect. 2 describes the method that supported the ontology development for both vertical and horizontal integrations; Sect. 3 presents the approach for the ontology-based data access; Sect. 4 describes the designed interoperability platform; Sect. 5 presents the PRODUTECH-SIF scenario and its asset efficiency tested; and finally, Sect. 6 presents the conclusions.

2   Ontology Development

We address semantic interoperability between systems supported by an ontology. Our ontology is based on Uschold and King Methodology that established a method that helps those interested in developing ontologies. The method encompasses four distinct phases [10]: identify the ontology purpose; build the ontology, that means capture the ontology, code the ontology, and integrate with existent ontologies if possible; evaluate if the ontology corresponds to the expected result; and finally, documentation that explains the main concepts of the ontology.

The ontology development aimed two separate models, one toward vertical integration and one toward horizontal integration. The vertical integration ontology uses the ISA 95 standard, and the horizontal integration ontology uses STEP standards. Both ontologies were modeled in the Protégé software tool.

For addressing the vertical integration (ISA 95-based) ontology, we divided our model into three sub-ontologies: hierarchy, operation type, and resource.

The hierarchy model refers to the breakdown structure of the involved actors in a process. Figure 1 depicts some of the classes of the model, and Table 1 depicts some of the ontology properties.

Fig. 1 Excerpt of the hierarchy sub-ontology classes

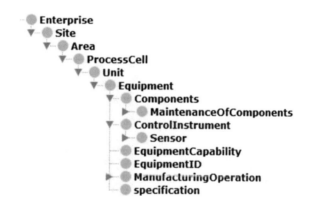

Table 1   Some hierarchy sub-ontology properties

Property type	Property name	Domain	Range
Object property	hasSite	Enterprise	Site
Object property	hasArea	Site	Area
Object property	hasProcessCell	Area	ProcessCell
Object property	hasUnit	ProcessCell	Unit
Object property	hasEquipment	Unit	Equipment
Data property	hasDescription	Owl:thing	Xsd:string
Data property	hasEquipmentID	Equipment	Xsd:string
Data property	hasEquipmentCapabilityType	EquipmentCapability	Xsd:string

The operation-type model refers to the breakdown of all tasks and jobs. This model aims to include semantics used in MES, since the model's domain is in line with typical MES' data models. Figure 2 depicts some of the classes of the model, as well as their properties in Table 2.

The resource model includes every element that is part of the production and manufacturing process. This model aims to include semantics used in ERP, since the model's domain is in line with typical ERP data models (workers, materials, machinery, etc.). Figure 3 depicts some of the classes of the model, and Table 3 depicts some of the ontology properties.

The horizontal integration was promoted by adopting STEP. STEP is the de facto standard for the information exchange between CAD/CAM/CAE systems. The objective was to transform the information into an ontology in order for the information to be more easily processed. To do that, we used a NIST plug-in for Protégé, called ontoSTEP [11], that allowed the transformation of a STEP file to an OWL file.

Fig. 2 Excerpt of the operation type sub-ontology classes

Table 2 Some operation type sub-ontology properties

Property type	Property name	Domain	Range
Object property	hasQuality	OperationType	Quality
Object property	hasproduction	OperationType	Production
Data property	hasJobListID	JobList	Xsd:string
Data property	hasPriority	JobOrder	Xsd:string
Data property	hasPublishedDate	WorkSchedule	Xsd:string
Data property	hasWorkMasterCapacityType	WorkMaster	Xsd:string
Data property	hasWorkPerformanceID	WorkPerformance	Xsd:string
Data property	hasWorkScheduleID	WorkSchedule	Xsd:string

Fig. 3 Excerpt of the
resource sub-ontology model
classes

Table 3 Some resource sub-ontology properties

Property type	Property name	Domain	Range
Object property	RequiresMaterialDefinition	ProcessSegment	MaterialResources
Object property	RequiresPersonnel	ProcessSegment	humanResources
Object property	RequiresPhysicalAsset	ProcessSegment	EquipmentResources
Data property	hasPhysicalAssetCapabilityType	Equipment	Xsd:string
Data property	hasPhysicalAssetID	Equipment	Xsd:string
Data property	hasPhysicalLocation	Equipment	Xsd:string
Data property	hasVendorID	Equipment	Xsd:string

3 Ontology-Based Data Access Approach

This section describes a design approach for enabling software access to heterogeneous databases using an ontology model described in the previous section. Access
mechanism to databases is based on an ontology-based data access (OBDA) application. As depicted in Fig. 4, an application OBDA receives as input a SPARQL query,

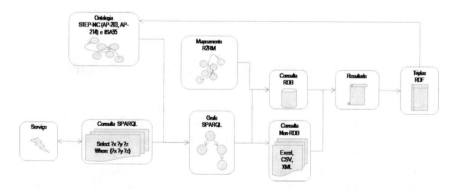

Fig. 4 Interoperability of all system

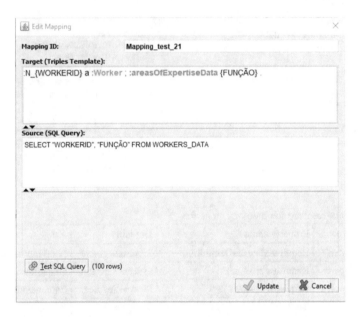

Fig. 5 Example of the ontology and the database mapping

then that SPARQL query is converted to a SQL query capable capturing the data and answering the question satisfactory.

In order to develop our OBDA, we used ONTOP. As all OBDA systems, ONTOP needs two things, a conceptual layer and a database. Our conceptual layer is both ontologies: the ontology that guarantees the vertical integration and the ontologies that guarantee the horizontal integration. The databases are from our ERP, MES, and Thingsboard. After having the conceptual layer and the database, we map one to another. The language used on ONTOP to the mapping is R2RML (RDB—relational database to RDF mapping language). The mapping between the ontology and the database was composed by mapping ID, source, and target (Fig. 5.). After the mapping phase, we can start using SPARQL queries on our databases.

4   The Cloud Collaborative Manufacturing Architecture

Industrial and manufacturing organizations are part of enterprise networks that work together, structuring themselves in product development flow activities. Efficiency of the flow is thus promoted by a harmonized collaboration between the enterprises, rather than each one working in a silo.

It is thus crucial that modern enterprise networks take advantage from existing technological infrastructures for orchestrating such collaboration. Cloud computing solutions have enabled exchange of process information through services that execute on the web. Namely, such services rely in protocols such as Application Programming Interfaces (APIs) for real-time communication.

Architecture design must include taking decisions on the orchestration of the collaboration within the enterprise network, the product development process, and the communication requirements (this one more related with the systems involved in each of the enterprises).

The orchestration of the collaboration is promoted by developing a set of services responsible for connecting different enterprises, where typically a set of APIs assure the information flow. The OBDA solution, proposed in the previous section, is included in such services, requiring an API for it as well. The API allows any service to query the existing SPARQL services included and hence use it as a service for the semantic interoperability between the enterprises.

The product development process must be addressed in the cloud architecture by developing a set of domain-oriented services, capable of managing information regarding different manufacturing scenarios. Other services like gateways, brokers, security, and data integrity may be included as well, as best practices for orchestration of the services.

Finally, defining needs for communication relies in the different existing layers within the enterprise. For this matter, industrial reference models like *Industrial Internet Reference Architecture* (IIRA) or *Industrie 4.0 Reference Architecture Model* (RAMI 4.0) propose division of layers, like enterprise (ERP, MES, and other business users), platform (cloud management), and edge (devices and assets). Between these layers, communication typically relies in protocols such as OPC-UA, MQTT, AMQP, or HTTP.

5   Collaborative Manufacturing in the PRODUTECH-SIF

In the project scope, our mission was to guarantee the interoperability between different hierarchics of an enterprise. Enterprises are dealing with a panoply of software's from different software houses, which produces different information types.

For example, an enterprise resource planning (ERP) usually does not show any distinction between workers and equipment's. What we really are trying to say is that for an ERP, both are nothing more than resource. On the other hand, for a manufacturing execution system (MES), normally, an equipment is a "machine", and the term "personnel" refers to a "human resource". The loss of semantic information is addressed using ontologies, as described in Sect. 3.

In this research, an ERP is considered a centralized system that facilitates the exchange of information between different enterprise systems, while a MES is a system that monitors and manages all productions. MES will be seen as in-between from the ERP responsible for taking the decisions and the shop floor the place where things are actually manufactured.

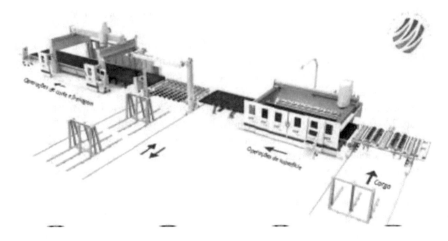

Fig. 6 Industrial scenario

Our objective is to integrate all the manufacturing system, vertically and horizontally, from the shop floor to the ERP; in other words, what we are trying to create is a cyber-physical system.

As a scenario, we are going to use a factory responsible for the manufacturing of tabletops, tombstones, and other types of stone products. As depicted in Fig. 6, generally, we can say that type of factory possesses two kind of machines, on one hand, you have three-axis CNC machine responsible for the polishing of the stone, and a second one, five-axis CNC machine responsible for the cutting of the stone. Associated with them, the factory also has all sort of other equipment.

We want to guarantee the factory interoperability from the shop floor to the ERP. In our scenario, the ERP system was developed by *Vanguarda Soluções De Gestão E Contabilidade Empresarial, Lda.*, and the MES system was developed by *INOCAM Soluções de Manufactura Assistida por Computador, Lda.* Additionally, computerized numerical control (CNC) machines were manufactured by *Companhia De Equipamentos Industriais, Lda.* (CEI), part from the *Zipor group.* The data is going to be captured thanks to the new Internet of Things (IoT) technologies and then sent to an IoT platform, namely Thingsboard.

The asset efficiency (AE) testbed in the PRODUTECH-SIF project was designed, so operational information from equipment's in a shop floor—in this case, only from a CNC—could be analyzed from users inside and outside of the enterprise. The analyzed data included working hours, temperature, energy consumption, and vibration. Additionally, process data was included as well. Bills of materials, warehouse stocking materials, and production orders were gathered from the Vanguarda's ERP. Production operations and control data were gathered from INOCAM's MES.

Following trends such as product lifecycle digital thread and digital twins, equipment's, materials, and processes were modeled in an ontological representation (OWL), aggregating ISA-95 and STEP (AP-203, AP-214). The ontology was able to be queried by means of an API, which enabled other services gaining access to the SPARQL queries.

Finally, data visualization in the Thingsboard platform was performed through acquisition of the data from the CNC, by means of the configuration of telemetry analysis services using MQTT, HTTP protocols. Thingsboard platform includes data analysis services like dashboards, which were used to monitor the AE.

Now that the operational data is available in the cloud (i.e., the Thingsboard platform), that data is able for usage in the business perspective. The services deployed in Thingsboard that promoted cloud collaboration rely mainly in business configurations—customers, users, devices, business rules.

6 Conclusions and Future Work

This paper presented the results of an ontology development for achieving the semantic interoperability. One of the results is an OWL based on ISO10303 and ISA-95. The adoption of these standards promotes a common data model that a widespread number of heterogeneity systems could relate to and communicate with. These data models now possess meaning, whereas materials relate to the capabilities included in their industrial digital twin model. STEP covers a wide range of products (electronic, electromechanical, mechanical) and stages of product development (design, analysis, manufacturing). On the other side, the data model also possesses meaning, whereas process monitoring and control are traced within the industrial digital thread model. ISA-95 was used for the concepts relating to interoperability between ERP, MES, and the shop floor systems. Both standards are widely recognized for application and product lifecycles, respectively.

Then, an OBDA-based approach was implemented for allowing different systems to interoperate using an implemented API that allows access for external services to the SPARQL queries. It was used as one of the services within the cloud collaborative manufacturing architecture. Other services aimed at connecting enterprises and acquiring shop floor data from a CNC to the cloud, to be visualized in a Thingsboard platform. Based on an interoperable scenario, we have the objective to develop an asset efficiency, but it still needs to be deployed and tested in a real shop floor. As the future research, it is still needed to address concerns regarding acquisition of material data, access to CNC machines, acquisition of sensors, among others.

Acknowledgements This work was carried out within the scope of the project "PRODUTECH SIF-Soluções para a Indústria do Futuro" reference POCI-01-0247-FEDER-024541, co-funded by Fundo Europeu de Desenvolvimento Regional (FEDER), through Programa Operacional Competitividade e Internacionalização (POCI), and by FCT—Fundação para a Ciência e Tecnologia within the R&D Units Project Scope: UIDB/00319/2020.

References

1. Jay Lee, B. B. H.-A. K. (2014). A cyber-physical systems architecture for industry 4.0-based manufacturing systems. *Manufacturing Letters*.
2. Jianhua Shi, J. W. H. Y. H. S. (2011). A survey of cyber-physical systems. In *Proceedings of the Inernational Conference on Wireless Communications and Signal Processing*.
3. Tursi, A. (2009). Ontology-based approach for product-driven interoperability of enterprise production systems.
4. Hřebíček, J. Semantic Interoperability. [Online]. Available: http://www.iba.muni.cz/obr/File/organize/semanticka_interoperabilita.pps.
5. Chandrasekaran, B. J. J. B. V. R. (1999). What are ontologies, and why do we need them?, *Intelligent Systems and their Applications, IEEE*.
6. Lin, D. S.-W. (2017). Industrial IoT/industrie 4.0 viewpoints. [Online]. Available: https://industrial-iot.com/2017/09/iiot-smart-manufacturing-part-2-digital-thread-digitaltwin/. Accessed May 2018
7. ISO. (2003). ISO/DIS 10303–238. Industrial automation systems and integration—product data representation and exchange—part 238: application protocols: application interpreted model for computerized numerical controllers. ISO.
8. ISA-95. (2000). ANSI/ISA-95.00.01–2000, enterprise-control system integration part 1: models and terminology. ANSI/ISA.
9. Calvanese, D. G. G. L. D. E. A. (2009). Ontologies and databases: The DL-lite approach. In *Reasoning Web. Semantic Technologies for Information Systems* (pp. 255–356). Springer.
10. Uschold, M., & King, M. (1995) Towards a methodology for building ontologies. In *Workshop on basic ontological issues in knowledge sharing*.
11. Raphael Barbau, S., Krima, S., Rachuri, A., Narayanan, X., Fiorentini, S. F., & Sriram, R. D. (2012). OntoSTEP: enriching product model data using ontologies.

Digital Platforms

A Benchmarking of Reference Models for Digital Manufacturing Platforms

Francisco Fraile, Víctor Anaya, Raquel Sanchis, Ángel Ortiz, and Raúl Poler

Abstract This paper presents a benchmarking of different reference models for Industry 4.0 solutions, using available alignment reports as a tool for benchmarking, a qualitative indicator to assess the appropriateness of the use of the different reference models, and an assessment using existing implementations and proposals as an initial starting point for future benchmark use cases. The main objective of the benchmark is to facilitate the adoption of reference models for the architectural definition of new digital manufacturing platforms. With this purpose, the benchmark first identifies the main synergies and complementarities of the different reference models under analysis and later performs a qualitative analysis of the relevance of the definitions they contain in the context of concrete implementations and proposals. In early stages of the definition of a new digital manufacturing platform, this is a useful start to position the proposal in the problem space spanned by the reference models and understand which aspects are really needed. The benchmarking can also be useful for the definition of new reference models for specific application domains or meta-models of reference models that aim to map features of different reference models in a common framework.

Keywords Cloud manufacturing · Methods and tools for interoperability · Enterprise application integration · Reference ontologies and standardization

1 Introduction

Industrial Internet of Things (IIoT) [1] integrates different technologies to collect product or process data originated in production environments, store these data, and gain insights through advance analytics, accurate predictions using machine learning or simulation capabilities implementing the digital twin pattern. IIoT is a fundamental part of digital manufacturing platforms [2], which leverage such services to support manufacturing in a broad sense, from product or process design to manufacturing

F. Fraile (✉) · V. Anaya · R. Sanchis · Á. Ortiz · R. Poler
Universitat Politècnica de Valencia, 46021 Valencia, ES, Spain
e-mail: ffraile@cigip.upv.es

B. Archimède et al. (eds.), *Enterprise Interoperability IX*, Proceedings of the I-ESA
Conferences 10, https://doi.org/10.1007/978-3-030-90387-9_9

Table 1 Reference model foundations

Model	Provided by	Based on
RAMI 4.0	Industrie 4.0 consortium	CIMOSA [8], SGAM [9], ISA-95 [10], IEC 62264 [11], IEC 62890 [12], OPC UA [13], AutomationML [14], AASX [15]
SMS	National institute of standards and technologies	SCOR [16], ISA-95 [10], CAM-I [17], CIMOSA [8], ATHENA [18], MTConnect [19], HTTP [20]
IIRA	Industrial internet consortium	ISO 42010 [21], BMM [22]
IMSA	Ministry of industry and information technology of China	CIMOSA [8]

operations. There is a great interest in the adoption of these services in the manufacturing industry. As a consequence, there is a growing number of digital manufacturing platforms and use cases that have emerged in recent years. The rapid advancement of related technologies (e.g., fields like big data, machine-to-machine communications, or data analytics) is another important factor that drives the appearance and evolution of digital manufacturing platforms.

Reference models provide a framework for the definition of complex systems and their related use cases. This common framework facilitates the architectural definition of the system and encourages standardization and interoperability. As described in [3], there are different reference models specifically designed for Industrial IoT systems and digital manufacturing platforms. The most prominent ones are the Reference Model for Industrie 4.0 (RAMI 4.0) [4], the Smart Manufacturing Standardization (SMS) Reference Model [5], the Intelligent Manufacturing Standardization Reference Model (IMSA) [6], and the Industrial Internet Reference Model (IIRA) [7]. Table 1 summarizes the main foundational models and standards in which the different reference models are based on.

The table highlights that although they all have similar objectives and there are synergies between them, they are different in scope, are based on different sets of standards, and provide somewhat overlapping definitions. These facts underpin the main objectives of this research paper: (a) map the different reference models against each other and conform a space where concrete implementations can be placed to better understand what aspects are relevant and (b) assess the relevance of the definitions in this space in the context of existing implementations and outstanding proposals to provide a useful starting point for new platform-related projects.

2 Benchmarking Methodology

The first step of the methodology is to align the definitions in the different reference models so that they can be evaluated in a meaningful way. The alignment used in this research paper is based on existing alignment reports in [4, 23, 24]. Based on these results, it is possible to use the four architectural viewpoints defined in IIRA, the business viewpoint, the usage viewpoint, the functional viewpoint, and the implementation viewpoint as four base dimensions for the alignment. This way, the RAMI 4.0 life cycle dimension and the RAMI 4.0 value streams can be mapped to the usage dimension, the IMSA life cycle, and NIST perspectives fit in the usage dimension. Likewise, the RAMI 4.0 layers and hierarchical levels, NIST 300-5 layers and ISA-95 levels, and IMSA layers and hierarchical functions fit in the functional dimension. Finally, the RAMI 4.0 administration shell and connectivity, the NIST AMS 300-2 (manufacturing data), AMS 300-4 (wireless), and AMS 300-6 (blockchain) fit into the implementation viewpoint (Fig. 1).

Based on this alignment, it is possible to perform an independent qualitative assessment to analyze and compare the different (alternative) definitions and determine to which extent they are relevant in the context of a concrete proposal and its related use cases. In this paper, six commercial platforms and research projects in digital manufacturing have been selected for the assessment. The benchmark indicator is a qualitative measure of the relevance of each definition for each implementation or proposal. To obtain this measure, first, a group of experts rated the relevance of each definition in each reference model in a scale from 1 to 10. Then, the average score is calculated, and the benchmark indicator is expressed as one of the following categories: ✓—relevant (10–7 score), (✓)—relevant to some extent (7–4 score), and ✗—out of scope (4–1 score). The following section shows the percentage of definitions that fall into each category based on the alignment results.

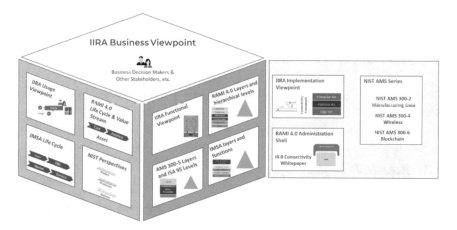

Fig. 1 Reference model alignment results

Table 2 Commercial platforms

Commercial platform	Platform provider	References
Mindsphere	Siemens	[25]
Thingworks	PTC	[26]
Predix	GE	[27]
IBM Cloud	IBM	[28]
Azure IoT Suite	Microsoft	[29]
Adamos	Software AG	[30]

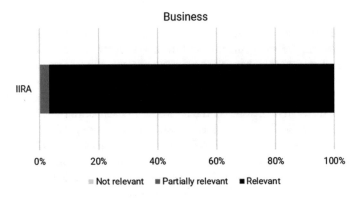

Fig. 2 Commercial platform benchmarking: Business viewpoint

3 Benchmarking Results

3.1 Commercial Platforms

Table 2 lists the different commercial platforms selected for the benchmarking (Figs. 2, 3, 4, and 5).

3.2 Research Projects

Table 3 lists the different research projects in digital manufacturing platforms selected for the benchmarking (Figs. 6, 7, 8 and 9).

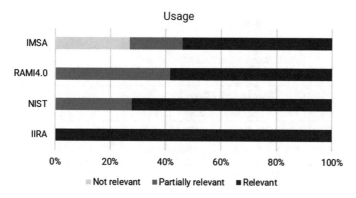

Fig. 3 Commercial platform benchmarking: Usage viewpoint

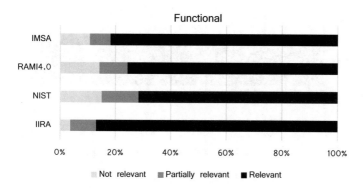

Fig. 4 Commercial platform benchmarking: Functional viewpoint

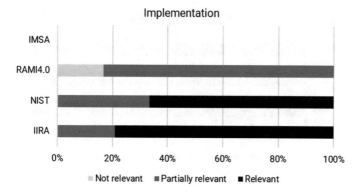

Fig. 5 Commercial platform benchmarking: Implementation viewpoint

Table 3 Research projects platforms

Research project	Title	References
ZDMP	Zero Defects Manufacturing Platform	[31, 32]
vf-OS	Virtual Factory Open Operating System	[33, 34]
CREMA	Cloud-Based Rapid Elastic Manufacturing	[35, 36]
C2NET	Cloud Collaborative Manufacturing Networks	[37, 38]
FIWARE	Future Internet Core Platform	[39, 40]
QU4LITY	Certifiable and Highly Standardized, SME-Friendly, and Transformative Shared Data-Driven ZDM Product and Service Model for Factory 4.0	[41, 42]

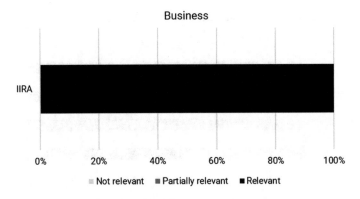

Fig. 6 Research project benchmarking: Business viewpoint

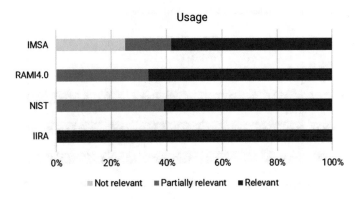

Fig. 7 Research project benchmarking: Usage viewpoint

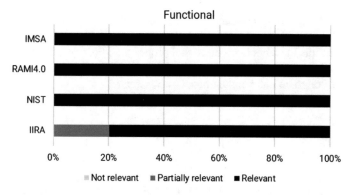

Fig. 8 Research project benchmarking: Functional viewpoint

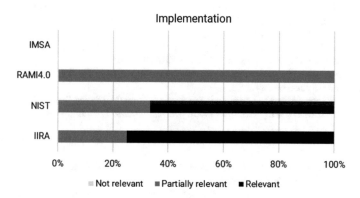

Fig. 9 Research project benchmarking: Implementation viewpoint

4 Conclusion

The assessment shown in this paper provides researchers and practitioners with a good starting point about the coverage of each reference model using existing implementations and proposals as an example. This will support them in the decision-making process about which reference model fits better for their specific project.

The main improvements that can be introduced in future research works are related to the number of reference models covered. Other reference models could be incorporated into the framework, first aligning them to the reference model alignment and then conducting the computing the qualitative measure of relevance with a group of experts. The incorporation of new reference models could also result in the definition of additional dimensions gathering for instance sustainability aspects, so as to define additional perspectives to assess the relevance of the reference models.

Finally, the assessment conducted has not been validated nor analyzed in detailed. The objective is to serve as example for other proposals, and due to the limitations in length, the results have not been discussed properly. In lines of this, future research should consider an in-depth analysis and validation of the assessment results, possibly conducted through an independent panel of experts.

References

1. Sadeghi, A. R., Wachsmann, C., & Waidner, M. (2015, June). Security and privacy challenges in industrial internet of things. In *2015 52nd ACM/EDAC/IEEE design automation conference (DAC)* (pp. 1–6). IEEE.
2. European Factories of the Future Research Association. (2016). *Factories 4.0 and beyond.* Working Document, Recommendations for the Work Programme, (pp. 18–19).
3. Fraile, F., Sanchis, R., Poler, R., & Ortiz, A. (2019). Reference models for digital manufacturing platforms. *Applied Sciences, 9*(20), 4433.
4. Deutsches Institut für Normung e. V., "Reference Architecture Model Industrie 4.0 (RAMI 4.0) English translation of DIN SPEC 91345:2016-04.," 2019.
5. American National Institute of Standards and Technology. (2016). *Current standards landscape for smart manufacturing systems.*
6. Ministry of Industry and Information technology of China (MIIT) and Standardization Administration of China. (2015). *National intelligent manufacturing standard system construction guidelines.*
7. Industrial Internet Consortium. (2017). *The industrial internet of things volume G1: Reference architecture.*
8. AMICE Consortium. (1989). *Open system architecture for CIM.* Research Report of ESPRIT Project 688, Vol. 1, Springer-Verlag.
9. Trefke, J., Rohjans, S., Uslar, M., Lehnhoff, S., Nordstrom, L., & Saleem, A. (2013). Smart grid architecture model use case management in a large European smart grid project. In *2013 4th IEEE/PES Innovative smart grid technologies Europe ISGT Europe* (No. 978, pp. 1–5).
10. American National Standard. (2000). *ANSI/ISA–95.00.01–2000 Formerly ANSI/ISA–S95.00.01–2000.* Enterprise-Control System Integration Part 1: Models and Terminology.
11. International Electrotechnical Commission. (2016). *Enterprise-control system integration—Part 3: Activity models of manufacturing operations management.*
12. International Electrotechnical Commission. (2017). *Life-cycle management for systems and products used in industrial-process measurement, control and automation.*
13. OPC Foundation. (2017). *OPC UA specification: Part 1—concepts. Version 1.04, November 22.*
14. Lüder, N., & Schmidt, A. (2016). "AutomationML in a Nutshell.," Handb. Ind. 4.0 Produktion, Autom. und Logistik, 1–46., pp. 1–46, 2016.
15. German Federal Ministry of Economic Affairs and Energy. (2018). *Details of the Administration Shell: The exchange of information between the partners in the value chain of Industrie 4.0 (Version 1.0)*
16. Stewart, G. (1997). Supply-chain operations reference model (SCOR): The first cross-industry framework for integrated supply-chain management. *Logistics Information Management, 10*(2), 62–67.
17. Ferreira, P. M., Lu, S. C. -Y., & Zhu, X. (1990). *Conceptual model for process planning.* Consortium for Advanced Manufacturing International (CAM-I), Arlington, Texas.
18. Berre, A., Elvesæter, B., Figay, N., Guglielmina, C., Johnsen, S., Karlsen, D., Knothe, T., & Lippe, S. (2007). *The ATHENA interoperability framework* (pp. 569–580). Enterprise Interoperability II.

19. MTConnect Institute (2014, Last Accessed June 2017) MTConnect Standard, Version 1.3, Standard. http://www.mtconnect.org/standard-documents
20. World Wide Web Consortium (2011, Last Accessed June 2017) REST, Web Page. http://www.w3.org/2001/sw/wiki/REST
21. Industrial Standards Organisation. (2011). *ISO/IEC/IEEE: 42010:2011 systems and software engineering—architecture description.*
22. Object Management Group. (2015). *Business motivation model (BMM).* http://www.omg.org/spec/BMM/
23. German Federal Ministry of Economic Affairs and Energy. (2018). *Alignment report for reference architectural model for industrie 4.0/intelligent manufacturing system architecture.*
24. Industrial Internet Consortium. (2017). *Architecture alignment and interoperability: An industrial internet consortium and platform industrie 4.0 joint whitepaper.*
25. Mindsphere (2019, Last Accessed October 2019). Web Page. https://siemens.mindsphere.io/en
26. Thingworks (2019, Last Accessed November 2019). Web Page. https://developer.thingworx.com/en
27. Predix (2019, Last Accessed November 2019). Web Page. https://www.predix.io/
28. IBM Cloud (2019, Last Accessed November 2019). Web Page. https://www.ibm.com/us-en/marketplace/cloud-platform/resources
29. Azure IoT Suite (2017, Last Accessed October 2019). Web Page. https://azure.microsoft.com/
30. Adamos (2019, Last Accessed November 2019). Web Page. http://adamos.com/en
31. Zero Defects Manufacturing Platform, ZDMP (2019, Last Accessed October 2019). Web Page. https://www.zdmp.eu/
32. Zero Defects Manufacturing Platform, CORDIS (2019, Last Accessed October 2019). Web Page. https://cordis.europa.eu/project/rcn/219920/factsheet/enVirtual Factory
33. Open Operating System, vf-OS, (2016, Last Accessed October 2019). Web Page. https://www.vf-os.eu/
34. Virtual Factory Open Operating System, CORDIS, (2016, Last Accessed October 2019). Web Page. https://cordis.europa.eu/project/rcn/205550/factsheet/en
35. Cloud-based Rapid Elastic Manufacturing, CREMA, (2015, Last Accessed October 2019). Web Page. https://www.crema-project.eu/
36. Cloud-based Rapid Elastic Manufacturing, CORDIS, (2015, Last Accessed October 2019). Web Page. https://cordis.europa.eu/project/rcn/193459/factsheet/en
37. Cloud Collaborative Manufacturing Networks, C2NET, (2015, Last Accessed October 2019). Web Page. http://c2net-project.eu/
38. Cloud Collaborative Manufacturing Networks, CORDIS, (2015, Last Accessed October 2019). Web Page. https://cordis.europa.eu/project/rcn/193440/factsheet/en
39. Future Internet Core Platform, FI-WARE, (2011, Last Accessed October 2019). Web Page. https://www.fiware.org/
40. Future Internet Core Platform, CORDIS, (2011, Last Accessed October 2019). Web Page. https://cordis.europa.eu/project/rcn/99929/factsheet/en
41. Digital Reality in Zero Defect Manufacturing, QU4LITY, (2019, Last Accessed January 2020). Web Page. https://qu4lity-project.eu/
42. Sesana, M., & Moussa, A. (2019). Collaborative augmented worker and artificial intelligence in zero defect manufacturing environment. In *MATEC web of conferences* (Vol. 304, p. 04003). EDP Sciences.

A B2B Marketplace eCommerce Platform Approach Integrating Purchasing and Transport Processes

Suat Gönül, Doğukan Çavdaroğlu, Yildiray Kabak, Dietmar Glachs, Fernando Gigante, and Quan Deng

Abstract Business-to-business (B2B) marketplace eCommerce platforms have grown in number in the last years. While these platforms allow product/service discovery and purchasing, they are limited in terms of integrating transport processes via well-defined B2B interactions. We approach this problem from an holistic view by dividing it into four sub-problems: enriching product and service descriptions with adequate semantic annotations for smooth discovery; integration of product classification taxonomies and standardized supply chain data representations; integration between purchasing and transport processes at procedural and data model levels; and disconnection between eCommerce platforms and legacy information systems of platform users. In this paper, we proposed a solution for each of these problems and presented a unified approach to integrate purchasing and transport phases in B2B marketplace eCommerce platforms. Finally, we validated the proposed approach with a case study in the scope of the NIMBLE research project including integration of eClass and Furniture Sector Taxonomy classification taxonomies into NIMBLE, semantic annotation of products with the information embedded in those taxonomies and a B2B scenario covering purchasing and transport phases of a traditional supply chain.

S. Gönül (✉) · D. Çavdaroğlu · Y. Kabak
SRDC Yazilim Arastirma, Gelistirme ve Danismanlik Anon. Sti, Ankara, Turkey
e-mail: suat@srdc.com.tr

D. Çavdaroğlu
Department of Computer Engineering, Middle East Technical University, Ankara, Turkey

D. Glachs
Salzburg Research, Jakob Haringer Street 5/II, 5020 Salzburg, Austria

F. Gigante
AIDIMME, Benjamin Franklin Street 13, 46980 Paterna(Valencia), Spain

Q. Deng
BIBA-Bremer Institut für Produktion und Logistik GmbH, Hochschulring 20, 28359 Bremen, Germany

University of Bremen, Bibliothekstraße 1, 28359 Bremen, Germany

B. Archimède et al. (eds.), *Enterprise Interoperability IX*, Proceedings of the I-ESA
Conferences 10, https://doi.org/10.1007/978-3-030-90387-9_10

Keywords Interoperability · Taxonomies · Business process integration · B2B eCommerce platform · Supply chain

1 Introduction

B2B marketplace eCommerce platforms, with many-to-many modality where multiple suppliers and multiple buyers exist simultaneously on the same platform (as opposed to Direct B2B eCommerce platforms with one-to-many modality), have flourished in the last decade as they provide supplier participants with increased visibility and customer access and buyer participants with the ability to discover product/supplier alternatives and compare them. For both sides, such platforms facilitate communication and reduce transaction costs, help establishing trust and eventually accelerate overall supply chain activities [1, 2].

Considering the relationship between a buyer and a supplier, purchasing and transport are two main phases in a supply chain through which several B2B interactions such as information inquiry, quotation or operation planning take place. In this sense, B2B marketplace eCommerce platforms mostly offer functionalities for requesting quotation and ordering in relation to the purchasing phase. However, shipment and transport options remain usually limited to a few alternative delivery types like express or regular options; it is not possible for trading companies to agree on detailed delivery/shipment terms in a structured manner following the purchase activity.

We argue the following additional challenges towards an effective (leading successful trading activities) and efficient (reducing B2B interaction efforts) platforms. First of all, for better integration of purchasing and transport processes on a platform, products and services (will be referring these two concepts collectively as products from now on) must have well-defined representations that are also linked with the information models used in supply chain activities. Product representations must also have adequate semantic annotations so that users can discover them on the platform easily. Finally, companies' legacy systems must be kept synchronized with the activities performed on the marketplace, e.g. in terms of updating inventory status or order records.

Addressing the challenges above, main contributions of this study can be summarized as follows: we first propose a product classification ontology. Second, we show how this product classification ontology is mapped to a Universal Business Language (UBL) [3] standard-based supply chain data model. Third, we show two B2B business process flows for purchasing and transport phases of supply chains, respectively, based on the individual business processes defined in UBL. And last, we present a configuration mechanism that can be applied at each step of B2B business process flows to realize the synchronization between the platform and legacy systems of the platform users. In the rest of this paper, in Sect. 2, we present related eCommerce platforms, product classification taxonomies and data representation standards for products and supply chain processes. In Sect. 3, we present the main contributions listed above. In Sect. 4, we present a case study implementing the proposed approach in NIMBLE [4], which is a cloud-based B2B marketplace eCommerce platform targeting European industry actors such as suppliers, manufacturers or service providers. NIMBLE is currently being developed in a European research project with the same name in the Factories of the Future (FoF) area. We conclude the study after discussing limitations of the study, innovative business models enabled by the proposed approach and future work.

2 Related Work

B2B marketplace eCommerce platforms have emerged in number and variety in recent years. In addition to global platforms like Alibaba,[1] Amazon Business[2] or TradeKey[3] providing any type of products as well as services, there have emerged regional and sectorial platforms like BeTimber[4] for timber trading only or wlw.at,[5] which is an Austrian-based eCommerce platform. Although these platforms vary in size, geography or targeted sector, they usually support a limited B2B interactivity. This can be summarized as publish and sell modality for suppliers and search and buy modality for buyers without any means to communicate via structured B2B transactions throughout the supply chain activities.

As a superset of supply chain activities that can be supported in B2B platforms, UBL and GS1 [5] initiatives define a set of supply chain activities including but not limited to tendering, quotation, ordering, fulfilment or transport execution plan along with data entities to be exchanged in B2B transactions throughout relevant activities.

UBL, GS1 and GoodRelations [6] are initiatives providing widely used standards for describing products' master data (basic product characteristics) and supply chain data (dynamic information related to any trading activity). All these standards provide a base data entity representing individual products or services and a set of generic properties that can be used to enrich the base representation. For instance, the base data entity in UBL is *Item* entity, which can be enriched with *ItemProperty* such that each *ItemProperty* can have multiple values in numeric, textual or quantity (number and unit pair) types. Similarly, GoodRelations ontology includes *ProductOrService* base entity which is a domain *qualitativeProductOrServiceProperty* and *quantitativeProductOrServiceProperty* properties. Although, GS1 has a base data entity to represent product or services, i.e. *Product* entity, it does not offer a generic property allowing enriching the base entity with arbitrary details but domain-specific properties like *textileMaterial*. However, as exemplified in the next paragraph, product

---

[1] https://fuwu.alibaba.com/gps/buyer.htm.

[2] https://www.amazon.com/b2b/info/amazon-business.

[3] https://www.tradekey.com/.

[4] https://betimber.com/.

[5] https://www.wlw.at/en/home.

classification taxonomies have much more coverage on diversity range of products and product properties. There exist several product classification taxonomies for thorough classification of products/services. Global Product Classification (GPC) [7], Google Product Taxonomy [8], eClass [9] and UNSPSC [10] are some of the product classification taxonomies. Varying in size and coverage, these taxonomies provide a classification hierarchy composed of thousands of product classes each of which can be associated with a set of properties, e.g. product class: *mechanical pencil* and product property: *ink type*. For instance, the latest version of the eClass taxonomy contains ~45 thousand of product classes and ~20 thousand product properties[6] collaboration a more. Our approach differentiates from existing works by bringing B2B data exchange and eCommerce paradigms together, thus providing a structured manner for product discovery and B2B transactions.

Neither product representation standards, nor supply chain standards, nor product classification taxonomies by themselves are enough for a seamless trading experience in a B2B marketplace eCommerce platform. All these three concepts must be available in an integrated manner as addressed in the next section.

3   Integrating Purchasing and Transport in Supply Chains

We divide the main challenge of seamless B2B interactivity covering purchasing and transport phases in supply chains into four sub-problems: (1) definition of a product classification ontology, (2) integration of product classification ontology into the supply chain information models, (3) linking purchasing and transport phases of supply chains and (4) synchronization with legacy information systems. Each sub-problem is addressed in the subsequent sections:

3.1   Product Classification Ontology

eCommerce platforms must be extensible with respect to integration of external product classification taxonomies. There are many existing taxonomies as some of them were mentioned in the related work section, nevertheless they might still be inadequate in terms of coverage of domain-specific variety. In fact, this is what we faced in NIMBLE regarding the semantic annotation of logistics services. As a solution, we defined a new taxonomy coding the knowledge required to annotate logistics services. Having multiple taxonomies, a generic product classification ontology was required to represent the structure of the external taxonomies and knowledge incorporated in them.

For a basic usage, a product classification ontology requires a hierarchy of product classes, property descriptions that can be assigned to classes and unit/value lists

---

[6] http://wiki.eclass.eu/wiki/The_Release_Process.

that can be assigned to properties. We utilize OWL [11] semantics for capturing the semantics incorporated in the classification taxonomies. OWL is a valuable technology since it has built-in constructs for specification of the basic taxonomy elements. Using OWL, a hierarchical structure can be established via *rdfs:subClassOf* property. The listing below shows how the *Chair* class is defined as a subclass of *Seat* class.

```
<owl:Class rdf:about="#Chair">
    <rdfs:subClassOf rdf:resource="#Seat"/>
</owl:Class>
```

Product properties with literal value ranges are connected to the class via datatype properties. Below, the definition of *hasCertificate* datatype property is given. It has a domain of *Chair* class and range of *string*. This means that a Chair instance might has Certificate property with string value.

```
<owl:DatatypeProperty rdf:about="#hasCertificate">
    <rdf:type rdf:resource="owl:FunctionalProperty"/>
    <rdfs:domain rdf:resource="#Chair"/>
    <rdfs:range rdf:resource="xsd:string"/>
    </rdfs:range>
</owl:DatatypeProperty>
```

A set of fixed coded values and quantities with a set of fixed units are associated to the product classes via object properties that are interpreted specially. The listing below shows an example of defining a quantity property for a product class. *hasEstimatedDeliveryTime* property has a domain of *Seat* and range of *nimble:QuantityType*. Nevertheless, the property is available for *Chair* because of the subsumption relationship formed by *rdfs:subClassOf* property. The property definition also refers to a unit list including the units that are allowed for this property. The limitation of OWL for this case is that it does not offer a suitable construct to define value or unit lists. To enable systematic interpretation of such knowledge (i.e. allowed values or units for particular properties), we defined dedicated ontological elements: *nimble:hasCode, nimble:hasCodeList, nimble:hasUnit* and *nimble:hasUnitList*.

Benefiting from the OWL constructs, we also let taxonomy designers to relate two products via object properties. The listing below shows how the *Seat* class is extended with *hasMaterial* property referring to other products of *Material* type. This indicates that a *Seat* instance must refer to a *Material* instance via the *hasMaterial* property.

```
<owl:ObjectProperty rdf:about="#hasEstimatedDeliveryTime">
    <rdf:type rdf:resource="owl:FunctionalProperty"/>
    <rdfs:domain rdf:resource="#Seat"/>
    <rdfs:range rdf:resource="nimble:QuantityType"/>
    <nimble:hasUnitList rdf:resource="#DeliveryTimeUnitList"/>
</owl:ObjectProperty>

<owl:NamedIndividual rdf:about="#DeliveryTimeUnitList">
    <rdf:type rdf:resource="nimble:UnitList"/>
    <nimble:hasCode                        rdf:datatype="xsd:normal-
izedString">day</nimble:hasCode>
    <nimble:hasCode                        rdf:datatype="xsd:normal-
izedString">hour</nimble:hasCode>
    <nimble:hasCode                        rdf:datatype="xsd:normal-
izedString">month</nimble:hasCode>
    <nimble:hasCode                        rdf:datatype="xsd:normal-
izedString">week</nimble:hasCode>
    <nimble:id    rdf:datatype="xsd:normalizedString">Delivery-
TimeUnitListId</nimble:id>
  </owl:NamedIndividual>
```

```
<owl:ObjectProperty rdf:about="#hasMaterial">
    <rdfs:domain rdf:resource="#Seat"/>
    <rdfs:range rdf:resource="#Material"/>
</owl:ObjectProperty>

<owl:Class rdf:about="#Material">
    <rdfs:subClassOf rdf:resource="#Product"/>
</owl:Class>
```

3.2   Mapping Category Data Model to UBL Data Model

This mapping approach is part of a larger data modelling effort as described in [12]. However, the initial study does not focus on definition of a generic product classification taxonomy.

As the base supply chain and product representation data model, we use Universal Business Language (UBL). In summary, UBL is used as the common data model for describing product/services details as well as the messages exchanged via the business processes. UBL's data model library contains reusable data entities in varying granularities. In this sense, UBL has also a good coverage of the concepts such as companies, persons, products, product properties, trading terms, clauses or contracts.

As mentioned in the related work, in UBL, products or services are represented with the *Item* data entity. A product can be provided with additional details via the *AdditionalItemProperty* entities and can be classified with a *CommodityClassification* entity, which in turn contains a coded value for the classification value. Figure 1 shows data structures for describing products with UBL and the proposed classification taxonomy as well as the mappings between these two models. A taxonomy

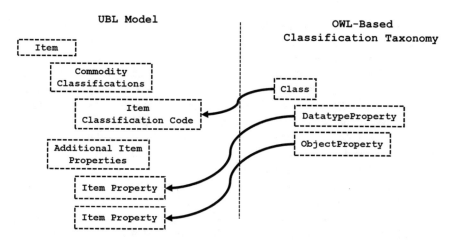

Fig. 1 Mapping knowledge from production classification taxonomies to UBL

class is mapped to the *ItemClassificationCode* entity of the *CommodityClassifica-tion* entity. In addition to the class mapping, each taxonomy class *property* (either datatype or object) is mapped to an individual *AdditionItemProperty*.

3.3   Connecting Ordering and Transport Phases

Although business processes usually represent complex flows in conventional usage, we restrict business processes to bilateral data exchanges composed of a request and a corresponding response between trading companies. Using such business processes, we introduce two business process flows addressing the activities on purchasing and transport phases in supply chains as depicted in Fig. 2.

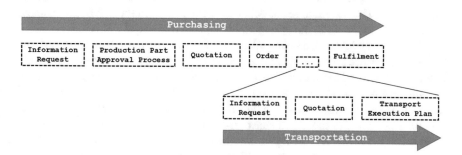

Fig. 2   Business process flows for purchasing and transport phases

The two flows represent interactions among three companies such that while the purchasing flow is composed of business processes between the buyer and supplier, transport flow is executed between the supplier and a transport service provider.

The label arrows show the direction of execution for each flow. Thus, the first two steps in the purchasing flow, which are *Information Request* and *production part approval process (PPAP)*,[7] can be classified as information inquiry steps where the buyer can request detailed information about the product of interest itself or production processes of the product, respectively. The information inquiry steps provide the supplier with the flexibility to decide on the level of information to be shared with the customer considering the confidentiality or sensitivity of the requested information. From the buyer's perspective, on the other hand, the inquiry steps facilitate trust forming towards the supplier as the revealed information would validate the supplier's promises about the product or production processes. Following the inquiry steps, trading companies can negotiate on the trading terms via the quotation step. In case of a successful negotiation, the flow continues with the order step. The next and last step in the purchasing flow is fulfilment. However, before proceeding with the last step, the supplier might optionally initiate a transport flow with a transport service provider with the eventual aim of organizing a transport service for shipping the ordered products to the agreed delivery address complying with the agreed delivery terms.

It is important to note that business process types included in these flows have been identified based on the requirements of use cases of the NIMBLE research project. We do not claim that the flows are complete in terms of the activities that can be performed in respective supply chain phases. For instance, UBL offers other processes such as tendering or billing that are not included in the proposed flows.

Most of the documents (messages) exchanged via the business processes are defined by the UBL standard such as UBL RequestForQuotation and Quotation documents are used in the quotation process, Order and OrderResponse in the order process and so on. We defined our own documents only for the PPAP process by following the convention of the standard, e.g. by adding similar mandatory fields like *ID* or list of *DocumentReferences* to refer other documents exchanged throughout supply chain. The complete set of documents exchanged in the NIMBLE business processes can be found in the open-source GitHub repository.[8]

At the data model level, the integration between purchasing and transport happens by instantiation of a transport-related business process using the information available at the order step of the purchasing flow. Documents used in the purchasing flow contain a list of *LineItem* entities referring to the product(s) of interest as well as the trading terms agreed throughout the flow for each product. As depicted in Fig. 3, the supplier has flexibility to initiate transport process(es) for combinations of products included in the order. According to the figure, the same transport service is being used for the first two products, but another service is arranged for the last one. *Item*

---

[7] https://www.aiag.org/quality/automotive-core-tools/ppap.

[8] https://github.com/nimble-platform/common/tree/master/data-model/ubl-data-model/src/main/schema/NIMBLE-UBL-2.1/maindoc.

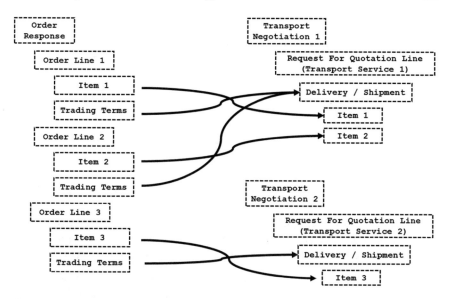

Fig. 3 Instantiating a transport quotation with order response

1 and *Item 2* configurations (as selected by the buyer) are mapped to the products to be shipped via *Transport Service 1*. Delivery-related trading terms included in the order response are also mapped to the delivery terms of the *Request for Quotation Line* of the transport service quotation.

Furthermore, to keep a connection between the transport flow and purchasing flow, we create a document reference from the request document of the first transport process to the response document of the order business process using the *AdditionalDocumentReference* construct, which is available in all UBL documents.

3.4 Configurable Business Processes

For a seamless data exchange between the trading companies by also keeping their legacy systems in sync with the activities performed on eCommerce platforms, we divide each step of a business process into three sequential tasks named *Document Creator*, *Document Processor* and *Document Sender*. As depicted in Fig. 4, both request and response steps of a business process are divided in this manner. The aim of each task can be summarized as follows:

- *Data Creator*: The message to be sent to target company is generated using own representation format of initiator company.
- *Data Processor*: The message generated in the *Data Creator* task may be stored based on the data management strategy of the initiator company, and it can

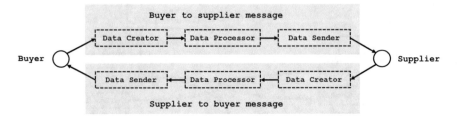

Fig. 4 Realization of a business process data exchange

be transformed into representation format of the target company via dedicated transformers.

- *Data Sender*: The message that can be consumed by the trading company is sent to a preconfigured endpoint. Access-control policies can also be applied in this step to ensure the privacy and security of the information included in the message.

As data sharing is subject to security and privacy concerns, it should be possible to skip the platform and send the data directly to the recipient. Targeting this requirement, we propose a *business process client* component that would contain company-specific logic for creating, processing and sending the message content to be implemented both for the buyer and supplier sides. From a deployment perspective, a business process client might be deployed both on the platform and in companies' premises. This approach provides sharing sensitive data directly with the trading company bypassing the platform as shown in Fig. 5. However, users may prefer platform to access the exchanged data on which value-added services, e.g. real-time monitoring, can be built as shown in Fig. 6. In either cases, clients would inform the platform so that the platform can track the status of the overall business process flow.

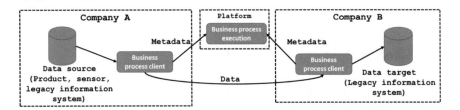

Fig. 5 Deployment topology for B2B data exchange bypassing the platform

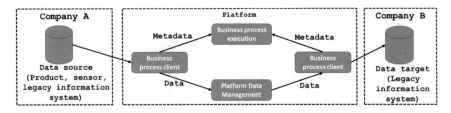

Fig. 6 Deployment topology for B2B data exchange via the platform

4 Case Studies

We present three case studies based on the NIMBLE research project (accompanying a B2B marketplace eCommerce platform addressing the four sub-problems addressed in Sect. 3). In the first case study, we explain how to integrate product classification taxonomies into NIMBLE and how to use them to semantically annotate and classify products. In the second one, we present a scenario covering purchasing phase. In this scenario, the supplier arranges the transport activities out of the platform after purchasing phase is completed in the NIMBLE platform. The third case study shows how to integrate both purchasing and transport phases in NIMBLE.

4.1 Case Study 1—Product/Service Discovery on NIMBLE

4.1.1 Integration of Classification Taxonomies into NIMBLE

We have integrated two external product classification taxonomies, namely eClass and Furniture Sector Taxonomy (FST) [13]. eClass is a cross-sector ISO/IEC compliant industry standard for product and service classification including thousands of product classes and associated properties. Despite being such a large taxonomy, it was inadequate in capturing furniture sector-related concepts. Furthermore, the coverage of the logistics service classification was also not sufficient for the furniture and eco-house use cases of NIMBLE. Therefore, we defined FST based on the ISO Standard for the Exchange of Furniture Product Data (funStep)[9] including concepts related to industrial processes, machinery, techniques, materials, components as well as products and product categories.

eClass is originally represented with a relational model.[10] We have transformed the relational model into the classification ontology structure proposed in Sect. 3.1. We defined the FST from scratch by using the proposed structure directly. Once taxonomy integrations were complete, we persisted them in a free-text index to be served to semantically annotate products while publishing them to the NIMBLE platform.

4.1.2 Classifying Products with the Integrated Taxonomies

Once the taxonomies are integrated, we were able to classify products published on NIMBLE with classes from both taxonomies. Figure 7 shows how a transport service is annotated with classes and properties from multiple taxonomies. After the annotation phase, products become discoverable on the platform with the knowledge integrated from the taxonomies.

---

[9] https://www.iso.org/organization/275604.html.

[10] http://wiki.eclass.eu/wiki/Category:Structure_and_structural_elements.

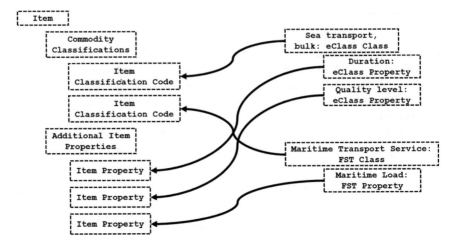

Fig. 7 Semantic annotation of transport service with knowledge from multiple taxonomies

Now, a supplier, who would like to use the services available on the platform for an incoming order, can initiate a transport flow on them. NIMBLE provides faceted and semantic search mechanisms where the search filters are dynamically generated based on both the classes and properties ingested from the classification taxonomies. Considering the example in Fig. 7, users will be able to filter search results by service class and duration, quality level and maritime load properties.

4.1.3   A Scenario Covering Product Discovery on NIMBLE

This scenario contains the following artificial companies:

- Company A is a Russian company searching for dining chairs with specific characteristics on NIMBLE.
- Company B is a Spanish supplier of such chairs in NIMBLE platform, and the scenario has the following sequence of activities:

  1. Company B publishes a product named "Wooden dining chair" by annotating it with "*Glue laminated timber*" eClass category and "*Chairs*" FST category and properties defined for those categories.
  2. Company A searches for *waterproof, glued, laminated dining chairs made of timber* on NIMBLE by using the corresponding search filters generated based on eClass and FST properties.
  3. Company A selects a chair named "Wooded dining chair" whose supplier is Company B among many alternatives.

4.2   Case Study 2—A B2B Scenario Covering Purchasing Phase

This scenario builds on the first scenario. After the purchasing is completed in NIMBLE, the supplier uses an external transport service provider to ship its products. The scenario has the following sequence of activities:

1. Company A initiates a purchasing flow with Company B resulting in ordering of the "*Wooden dining chair*" product. The flow might contain several sub-processes-related information inquiry (via Item Information Request and PPAP) and negotiation (via quotation). Negotiation step might be repeated until an agreement is reached.
2. Company B uses an external transport service provider to ship its products by initiating a fulfilment process with Company A.
3. Upon receiving the shipped products, Company A concludes the fulfilment process initiated by Company B.

4.3   Case Study 3—A B2B Scenario Covering Purchasing and Transport Phases

This scenario again builds on the first scenario. Compared to the previous case study, the supplier searches for a suitable transport service provider in NIMBLE and uses it to ship its products. Thus, we add the following company which represents a transport service provider in NIMBLE:

- Company C is a Spanish transport service provider which has a certificate to ship products from Spain to Russia.

  The scenario has following sequence of activities:
- Company A initiates a purchasing flow with Company B resulting in ordering of the "*Wooden dining chair*" product. The flow might contain several sub-processes-related information inquiry (via Item Information Request and PPAP) and negotiation (via quotation). Negotiation step might be repeated until an agreement is reached.
- Company B searches for a transport service provider providing a cheap service (probably a sea-based transport service) with "*Eco-Label certificate*" to ship the ordered products from Spain to Russia.

  (a)   Company B selects "*Sea transport door to door delivery service from Spain*" service from Company C out of the search results.

- Company B initiates a transport flow with Company C to ensure that the delivery planning complies with the delivery terms promised to Company A such as delivery period, incoterms, location or tax coverage. The flow results with the arrangement of the transport service for shipping the ordered dining chairs.

- Company B ships products using the transport service provided by Company C by initiating a fulfilment process with Company A.
- Upon receiving the shipped products, Company A concludes the fulfilment process initiated by Company B.

Figure 8 shows a summary of an example sequence of B2B interactions from the suppliers (Company B) point of view. As a supplier, the user is able to track the sequence of activities both with the buyer and with the transport service provider.

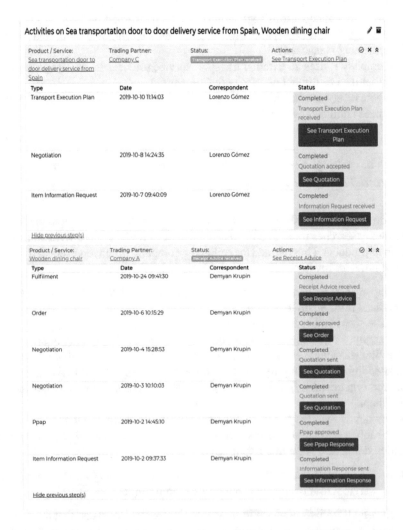

Fig. 8 Summary of B2B interactions of a supplier (Company B) with a buyer (Company A) and a transport service provider (Company C)

As the two sets of process flows are linked to each other, we were able to visualize them easily.

4.4 Configuration of Business Process Steps

In relation to the second and third case studies, we did not perform any integration with real legacy systems and therefore did not require any transformation between different data representation formats. Instead, the following default configurations are used:

- *Data Creator*: The messages to be sent to target trading company are generated based on the UBL standard by mainly using the user interface modules of the NIMBLE platform dedicated to visualizing the data for each business process step in the purchasing and transport flows.
- *Data Processor*: In this step, the generated message is persisted in a relational database in NIMBLE along with metadata containing information related to the business process instances that the message is related to. This indicates that the *business process clients* were included within the NIMBLE platform.
- *Data Sender*: As the complete supply chain data is managed in the scope of NIMBLE, this step is only used to notify the target trading company with the B2B activities happening.

5 Discussion, Limitations and Future Work

By nature, NIMBLE combines B2B data exchange and eCommerce concepts. In this respect, NIMBLE features (e.g. domain-specific and semantic knowledge-based discovery of products, progressive negotiation on strategic agreements or operational planning) enabled by the proposed approach go beyond *publish* → *sell* and *search* → *buy modalities* offered by the available eCommerce platforms for supplier and buyer users, respectively.

Trust is a critical factor for sustainability of eCommerce platforms [14]. Many eCommerce platforms including Alibaba take support from independent third-party organizations for validating their suppliers. As indicated by the end-users of NIMBLE in the furniture and eco-house sectors, it is even a frequent practice to visit suppliers' premises to validate the supplier, product and production processes. In this respect, Information Request and PPAP processes allow companies to establish trust as they allow acquiring more details about product and production processes, reflecting the identities and capabilities of the supplier.

Going beyond the integration of purchasing and transport flows, the business processes can be configured in more advanced ways such that a synchronization between production and transport flows can be achieved. In fact, NIMBLE already supports data channels (which are out of the scope of this study) through which production data are shared in real time. This capability provides NIMBLE users with just-in-time supply chain operations.

A limitation of this study is partial integration of knowledge from external classification taxonomies. For instance, in addition to the class hierarchy and properties, eClass taxonomy defines synonyms for classes and property labels. So, a future work would be to extend taxonomy model to represent such additional knowledge. Another limitation of the proposed approach is that the business process flows are composed of fixed set of business processes. It contains another extension that might be related to provide flexibility for defining new business processes on a platform as apposed to the fixed set of current processes offered to the users.

6 Conclusion

In this study, we proposed a B2B marketplace eCommerce platform approach for a seamless B2B interactivity covering both the purchasing and transport phases in a supply chain. We decomposed the overall problem into a set of sub-challenges, each of which is addressed individually. Addressing the first challenge, which is inadequate semantic annotation of products/services, we introduced a generic product classification taxonomy model and represented two taxonomies, namely eClass and Furniture Sector Taxonomy, with this model.

The second challenge we addressed was integration of the classification taxonomy model into the UBL data model, which is the supply chain data representation standard used in the NIMBLE research project encapsulating this study. We defined mappings between data entities that are used to represent products and their products.

Third challenge was to integrate purchasing and transport phases of supply chains. Addressing this challenge, we first defined dedicated workflows for these two phases representing, respectively, the sequence of B2B interactions for ordering/receiving a product and carrying the products from manufacturer to buyer. In addition to this high-level definition of B2B interaction flows, we presented the mechanism, at the data model level, to initiate a transport flow based on the information (i.e. details about the products to be transported, delivery-related terms agreed between the buyer and supplier) available within the purchasing workflow.

Lastly, we addressed the synchronization of legacy information systems of eCommerce platform users with the activities happening on the platform. We proposed a configuration mechanism for each data transmission step such that the creation of the message to be sent, transformation into other data representation formats, storage/transmission of the message to any endpoint and access-control rules can be configured. We presented a case study by exemplifying product publishing supported by the semantic annotation mechanisms thanks to the integration of external classification taxonomies. Organized around the published products, we defined a use case scenario including international purchasing and transport that can be realized with the proposed approach.

Acknowledgements This research has been funded by the European Commission within the H2020 project NIMBLE (Collaborative Network for Industry, Manufacturing, Business and Logistics in Europe), No. 723810, for the period between October 2016 and March 2020.

References

1. Baršauskas, P., Šarapovas, T., & Cvilikas, A. (2008). The evaluation of e-commerce impact on business efficiency. *Baltic Journal of Management, 3*(1), 71–91.
2. Garicano, L. & Kaplan, S. (2000). The effects of business-to-business E-commerce on transaction costs. *The Journal of Industrial Economics 49*. https://doi.org/10.2139/ssrn.252210
3. Universal Business Language Version 2.1 (2013, November 04). *OASIS standard*. http://docs.oasis-open.org/ubl/os-UBL-2.1/UBL-2.1.html
4. Innerbichler, J., Gonul, S., Damjanovic-Behrendt, V., Mandler, B., & Strohmeier, F. (2017, June). Nimble collaborative platform: Microservice architectural approach to federated IOT. In *2017 Global Internet of Things Summit (GIoTS)* (pp. 1–6). IEEE.
5. GS1 Global Office (2018). *What is GS1?* https://www.gs1.org/sites/default/files/docs/what_is_gs1.pdf. Last accessed 20 November 2019.
6. Hepp, M. (2008). GoodRelations: An ontology for describing products and services offers on the web. In *2008 Proceedings of the 16th International Conference on Knowledge Engineering and Knowledge Management*.
7. Gs1.at (2005). https://www.gs1.at/fileadmin/user_upload/GS1_Global_Product_Classification.en_lang.pdf. Last accessed 20 November 2019.
8. Google.com (2019). https://www.google.com/basepages/producttype/taxonomy.en-US.txt. Last accessed 20 November 2019.
9. Bondza, A., Eck, C., Heidel, R., Reigl, M. & Wenzel, D. (2018). *Toward smart manufacturing with data and semantics*. https://www.eclass.eu/fileadmin/downloads/ecl-Whitepaper_2018_EN.pdf. Last accessed 2019/11/20.
10. Xu, Y., Zou, S., Gu, A., Wei, L., & Zhou, T. (2012). Research on the complex network of the UNSPSC ontology. *Physics Procedia, 24*, 1863–1867.
11. Antoniou, G., & Harmelen, F. V. (2003). Web ontology language: OWL. *Handbook on Ontologies*.
12. Deng, Q., Gönül, S., Kabak, Y., Gessa, N., Glachs, D., Gigante-Valencia, F., Damjanovic-Behrendt, V., Hribernik, K., & Thoben, K. D. (2019). An ontology framework for multisided platform interoperability. In *Enterprise Interoperability VIII* (pp. 433–443). Springer.
13. Funstep.org (2019). http://funstep.org/furniture-sector-taxonomy/FurnitureSectorTaxonomy-v2.4.7.owl. Last accessed 29 November 2019.
14. Lee, S. J., Ahn, C., Song, K., & Ahn, H. (2018). Trust and distrust in e-commerce. *Sustainability, 10*(4), 1015.

Digital Twin

Digital Twin-Driven Design: A Framework to Enhance System Interoperability in the Era of Industry 4.0

Safaa Lebjioui, Mamadou Kaba Traoré, and Yves Ducq

Abstract Product development and manufacturing is entering a digital era, thanks to the progress made in data science and virtual technologies. The digital twin (DT) is one of the key concepts associated with this transition to Industry 4.0. Yet, in the literature, the term is differently used in various communities. In addition, the DT implementation in the product development process (PDP) lacks a conceptual ground, which hinders the proper use and wider application of this technology in engineering design and product life cycle management. This paper proposes an interoperability framework for digital twin-driven product design, based on data integration at different stages of the respective life cycles of the product and its digital twin. Such a framework can greatly help companies optimize their PDP.

Keywords Digital twin · Product life cycle · Product development process · Interoperability · Industry 4.0

1 Introduction

In recent years, the product development process (PDP) is becoming more digitalized than ever before. Although data has always been a relevant issue examined by different bodies of knowledge, it is now becoming an important asset in the industrial transformation. Data is gathered from various sources at different stages in a product life cycle.

Taking benefit of the progress made in data science and virtual technologies, the digital twin (DT) approach emerged as a data-based value chain, which is gradually becoming a key research trend in smart engineering. Yet, in the literature, the term

S. Lebjioui (✉) · M. Kaba Traoré · Y. Ducq
IMS Laboratory, University of Bordeaux, 351 Cours de la Libération A31, 33400 Talence, France
e-mail: safaa.lebjioui@u-bordeaux.fr

M. Kaba Traoré
e-mail: mamadou-kaba.traore@u-bordeaux.fr

Y. Ducq
e-mail: yves.ducq@u-bordeaux.fr

is used differently from one discipline to another. Moreover, the implementation of DT in the PDP lacks of a conceptual ground, which hinders the proper use and wider application of this technology in engineering design and product life cycle management. Driven by this need, this paper first reviews the concept of DT and its evolution. Then a reference model for digital twin-driven design is presented. On this basis, a framework to design a physical product and its DT by integrating different data from their respective life cycles is proposed. This can guide companies in how to deploy a DT and use the data it provides to support their PDP at different stages. A conclusion and future work are given at the end.

2   The Concept of Digital Twin

In this paper, we consider a DT as a digital model of a product, a system or a process, which includes all the data that can support different phases of the engineering activities. In this section, the several aspects of DT are briefly highlighted, and recent definitions are discussed.

2.1   Concept Definition

The concept of DT was first introduced by NASA in their integrated technology roadmap in 2003 [1]. The DT concept has initially been defined as a simulation model, which is paired with the system of interest in a way to continuously reflect changes happening in that system [1, 2], while in [3–5] it is defined as the set of digital information gathered, aggregated, and analyzed throughout the product life cycle (thus, it exists prior to the product design and after the product end of life). The pairing between the digital twin and the system of interest is based on data collected from the system, including historical data and sensor data.

Beside the modeling purposes, the digital twin concept has also been applied for other purposes, such as verification [5, 6], prediction [1], and analytic activities [7]. From the modeling and simulation perspective, it appears as a disruptive approach [8].

2.2   Concept Application

The concept of DT provides an effective way to learn more about smart manufacturing. Many studies have contributed to promote DT in industrial practice. The current applications of DT in industry are briefly summarized in this section.

Negri et al. [2] defined a DT as a virtual representation of a production system, both being synchronized to each other thanks to data sensed from smart devices.

The DT value chain includes mathematical models, which are used in Industry 4.0 engineering systems for forecasting and real-time optimization of the manufacturing processes.

Schleich et al. [9] explained that product life cycle management (PLM) developers focus on tightly coupling the physical product with its digital model in order to increase the industrial resilience and competiveness. For example, SIEMENS is deploying several Industry 4.0 concepts (including DT) in order to improve productivity and quality in manufacturing, while General Electric concentrates on DT-based predictions and performance evaluation of their products over a life span. TESLA's target is to develop a DT for every produced car, so that a concurrent data transmission between the car operating in the real-world and the plant can be ensured.

As a result of various existing DT definitions and industrial applications, multiple partial understandings of this concept can be found in the literature [8, 10, 11]. We see in data integration a generic way to gain a more general understanding. The next section is devoted to the levels of data integration.

3 DT-Driven Design Process

Empowered by a combination of different Internet of Things (IoT) technologies and interoperability, a DT can mirror the physical twin. It can also predict and address the potential issues. To do so, a range of sensors are imbedded with the physical model. These sensors transfer real-time data about the product and its environment. The data collected is then analyzed. The use of this feed of real-world data by the DT can enhance the data-driven design. Moreover, it can help manufacturers understand how products operate in the field and therefore enable companies to take more informed, market-driven decisions about future generation of products.

Accordingly, a DT can be considered as an enabler for information refinement and integration. By refining a large amount of data, the DT allows gaining useful insights that can guide design activities toward new design options. By integrating several types of data from various sources, the DT allows identifying hidden patterns and facilitating the cross-checking of data and information [11].

Figure 1 shows the DT-driven design process, with a standard data life cycle including data collection, data mining, data integration, data storage, data analysis, and data transmission. Using this life cycle, raw data and real-world data are transformed to valuable information that can be directly investigated by engineers to support their design options. The phases are described below:

Data collection: This is the primary stage. In DT-driven design, the real-time operating data is collected by sensors and integrated by data acquisition tools.
Data integration: Data integration consists in combining data which is located in different sources and providing users with an overall view.

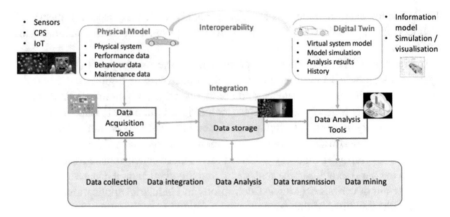

Fig. 1 DT-driven design process (D^3P): from data-driven to data-informed design

Data analysis: Data has always been at the center of decision making. Thus, this phase involves applying advanced data analytics tools to drive insights and find a correlation between the different data collected in order to turn it into information. *Data mining*: In DT-driven design, the purpose is to obtain information from the collected data set and transforming it into useful insight. This phase involves different steps such as data preprocessing, data management, physical and digital model inference, complexity consideration, visualization, and online updating [12]. Therefore, the DT-driven design can shift from being a data-driven design to a data-informed design. *Data transmission*: Data transmission refers to the process of transferring data between the physical product and its DT and vice versa. The interoperability of these two systems allows them to communicate between each other. The DT-informed design tries to assess the behavior that is behind the data; therefore, the real-time data sent by the physical product is a key input among other assets to build a deeper understanding of what value companies are providing to users.

The DT-driven design uses real-time data and becomes a data-informed design. Therefore, this can allow companies to make decisions which are grounded in reality and based on actual facts which can have a long-term downstream impact on the overall product development process.

4 Interoperability Framework for DT-Driven Design Process

In this paper, we consider a DT as a digital model of a physical asset. The digital model turns into a twin when it is connected to its physical component, system, or product. As shown in Fig. 2, there are three possible scenarios to develop a DT. The

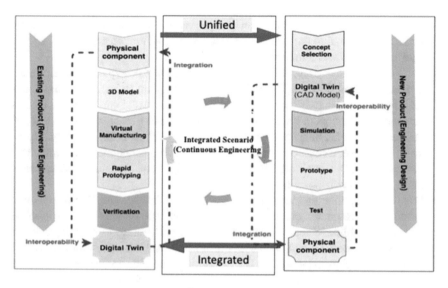

Fig. 2 Interoperability framework for D³P

different steps are going to be described in this section. However, it should be noticed that, in an industrial context, manufactures may not use the same steps to develop their DT. It is also possible to carry out these steps concurrently.

4.1 First Scenario: A DT of a New Product (DTNP)

- Process the available data to select a concept: The product development in the era of Industry 4.0 relies on companies' ability to handle data. Therefore, the first step is to process the empirical data. As discussed in the previous section (see Fig. 1), data can be collected from different sources. The collected data needs to be analyzed in order to be transformed into significant input that can be used by designers in order to select the right design concept for their product.
- Develop a DT model: At this stage, the commonly used technology in product design is computer-aided design (CAD). Designers develop these models using an existing database which provides different libraries that handle a wide range of production resource data, including bills of materials, layouts, interfaces, and other elements. To enable a holistic product/production view, companies must expand the functionalities of the commercially available software tools, with additional functionalities, i.e., pre-defined agents that contain additional technological information. Accordingly, the DT can easily interoperate with its physical twin at later stages.
- Simulate model/product behavior: The simulation is used to duplicate the key features and the expected behaviors of the physical product in the virtual world.

There are many types of simulation, like 3D motion simulation or discrete event that can be applied to the virtual model to ensure that the selected design will be able to meet the initial requirements.

- Build a prototype and test: Engineers usually develop prototypes, before a product is released for production. The prototypes help designers to learn more about how the product can be used and how it will operate in a real environment. Once a prototype is developed, it can be equipped with sensors and actuators to test and validate different features of the product before the design is implemented. On the other hand, new technologies can also be used, such as augmented reality. This can enable engineers to interact with the virtual product and test its functions in the simulated environment.
- Develop a physical product and connect it to its DT: When the development of a product is finished and connected to its DT, both become interoperable systems that have the ability to not only share information but to interpret incoming data. As the physical product is equipped with sensors that send back real-time operational data from the physical world, its DT can process this data and make adjustments to the physical product using the appropriate sensor technology and IoT.

Developing a product and its DT can give companies a behavioral outlook at any given point of the life cycle and enable continuous process adjustments.

4.2 Second Scenario: A DT of an Existing Product (DTEP)

Given an existing physical product:

- Do a reverse engineering: The principle is to reverse the engineering process of an existing component, system, or product, in order to develop a digital version. This process has two distinctive phases: The first one lies in collecting and digitalizing data, and the second one consists of creating a 3D model of the object based on the data collected. These data are fed to the step where the DT is created to enhance the interoperability and build a more functional DT.
- Simulation using virtual manufacturing: The concept of virtual manufacturing (VM) is widely used in literature, with a few definitions. VM makes use of virtual reality technologies to integrate diverse manufacturing-related technologies [13]. It provides a representation of the properties and behavior of a given product, using different analysis and simulation methods such as finite element analysis (FEA) and final volume method (FVM). Using this simulation makes possible for companies to optimize key factors which directly affect the product performance such as the final form and reliability in operation.
- Do a rapid prototyping before verification: Rapid prototyping involves different technologies, such as additive manufacturing or selective laser melting, which allow analyzing product functionality based on the physical model of a product.

- Connect the DT to its physical representation: While carrying out the reverse engineering, models are created and data is generated. The connection between the developed DT and its physical representation is enabled by various technologies.

The DT in this scenario can manage and control the data throughout the product life cycle. Accordingly, DT can carry out data collection, data transmission, and data storage from the real world.

4.3 Third Scenario: Integrated Scenario (Continuous Engineering)

The above scenarios are "one dimensional" and can be applicable if a product or a system is designed from scratch. However, nowadays, one of the main challenges is that product design is carried out incrementally and continuous changes are performed. Therefore, the design of the DT should not only focus on the "twining" between the physical and the virtual world but rather integrate the design activity of both.

The idea behind an integrated scenario is to expand and integrate the above scenarios to bring the fundamentals of interoperability and continuous engineering to the concept of DT.

Continuous engineering enabled by the integration of the physical and virtual world will allow engineers to handle the effects of the changes within the ecosystem where the products evolve. Unlike in the traditional V-model for systems engineering where the design is carried out in a sequential number of phases, in continuous engineering, activities are carried out iteratively across the PDP.

On the other hand, interoperability is a key feature in DT-driven design. An important notion in interoperability is the interaction between different systems. In this interaction, data can be federated, unified, or integrated. In the proposed framework, we assume that during the design of the product and its DT, whether this is done concurrently or following the traditional development steps, a number of agents interact by communicating information. This communication implies that one system exchanges and/or uses data of another one. Therefore, the interoperability issues that can arise from the data exchanged between a product and its twin is another challenge that needs to be considered in future work.

At its core, the concept of DT is all about how to increase the efficiency of product development through collaboration and system interoperability, using the operational data and real-time analytics to transform the gathered data into performance knowledge. This knowledge will help engineers to improve their engineering process and performance optimization.

5 Conclusion and Future Work

Modern PDP is not managed anymore by making assumptions about the product performance. Instead, performance-based analysis is used to provide real-world data on product performance from the field. Connecting the physical product to its DT thanks to the available IoT technologies will enable product to send usage data back to the platform and enhance their interoperability. The information gained from the real-world data provides a better understanding of field operations as well as useful insights for new business opportunities.

In this paper, we have proposed an interoperability framework for DT-driven product design. The framework can be seen as a first attempt to introduce a conceptual framework in that perspective, which need to be improved by future work. Thus, our next efforts will concentrate on the following aspects:

– Application of the proposed framework across a PDP to a real-world industrial case. This will help in identifying how the DT concept can concretely shape the PDP.
– DT life cycle management: As for a physical product, where PLM deals with all the data relative to a product across its life cycle, there is a need to consider the digital models and close the gap between product's physical and virtual spaces.

References

1. Boschert, S., & Rosen, R. (2016). Digital twin—The simulation aspect BT—Mechatronic futures: Challenges and solutions for mechatronic systems and their designers. In *Mechatronic futures*.
2. Glaessgen, E. H., & Stargel, D. S. (2012). The digital twin paradigm for future NASA and U.S. Air Force Vehicles. In *Collection of technical papers—AIAA/ASME/ASCE/AHS/ASC structures, structural dynamics and materials conference*.
3. Grieves, M., & Vickers, J. (2016). Digital twin: Mitigating unpredictable, undesirable emergent behavior in complex systems. In *Transdisciplinary perspectives on complex systems: New findings and approaches*.
4. Negri, E., Fumagalli, L., & Macchi, M. (2017). A review of the roles of digital twin in cps-based production systems. *Procedia Manufacturing, 11*.
5. Tuegel, E. J. (2012). The airframe digital twin: some challenges to realization. In *Collection of technical papers—AIAA/ASME/ASCE/AHS/ASC structures, structural dynamics and materials conference*.
6. Gockel, B. T., Tudor, A. W., Brandyberry, M. D., Penmetsa, R. C., & Tuegel, E. J. (2012). Challenges with structural life forecasting using realistic mission profiles. In *Collection of technical papers—AIAA/ASME/ASCE/AHS/ASC structures, structural dynamics and materials conference*.
7. Reifsnider, K., & Majumdar, P. (2013). Multiphysics stimulated simulation digital twin methods for fleet management. In *54th AIAA/ASME/ASCE/AHS/ASC structures, structural dynamics, and materials conference*.
8. Rosen, R., Von Wichert, G., Lo, G., & Bettenhausen, K. D. (2015). About the importance of autonomy and digital twins for the future of manufacturing. In *IFAC- PapersOnLine* (Vol. 28).

9. Schleich, B., Anwer, N., Mathieu, L., & Wartzack, S. (2017). Shaping the digital twin for design and production engineering. *CIRP Annals—Manufacturing Technology, 66*(1).

10. Lee, J., Lapira, E., Bagheri, B., & an Kao, H. (2013). Recent advances and trends in predictive manufacturing systems in big data environment. *Manufacturing Letters, 1*(1).

11. Tao, F., Cheng, J., Qi, Q., Zhang, M., Zhang, H., & Sui, F. (2018). Digital twin-driven product design, manufacturing and service with big data. *International Journal of Advanced Manufacturing Technology, 94*(9–12).

12. Chakrabarti, S., Ester, M., Fayyad, U., & Gehrke, J. (2006). Data mining curriculum: A proposal. In *ACM SIGKDD*.

13. Shukla, C., Vazquez, M., & Frank Chen, F. (1996). Virtual manufacturing: An overview. *Computers and Industrial Engineering, 31*(1–2).

A Digital Twin Model-Driven Architecture for Cyber-Physical and Human Systems

Milad Poursoltan, Mamadou Kaba Traore, Nathalie Pinède, and Bruno Vallespir

Abstract The cyber-physical and human system (CPHS) is widely recognized as a key infrastructure to support the future developments in healthcare, industrial manufacturing as well as in many other areas. In consequence, there is an increasing interest in tools, techniques, and technologies to advance the understanding and provide unchallenging improvement of CPHS. Digital Twin is a growing research topic that can equip the CPHS with high-fidelity mirroring, monitoring, controlling, and active functional improvement. The central question in this study asks in what way Digital Twin should be designed and developed for a CPHS. An architecture was required to answer this question and build upon model to define guidelines and requirements for valid Digital Twin. Thus, this study provides much new knowledge about the emerging role of Digital Twin in CPHS by using a model-driven architecture (MDA) approach with the perspective of SD logic. In addition, in the presence of human roles, the given MDA is beneficial for providing abstractions of Digital Twin from different viewpoints, communication with non-technical experts, and decision support, system design, and improvement.

Keywords Cyber-physical and human systems · Digital Twin · Model-driven architecture · SD logic

1 Introduction

CPHS comprises cyber, physical, and human components designed for controlling, monitoring, and improving through an integrated system. As a complex system, CPHS understanding is hard for humans, and intervention in such a system is even more difficult. Digital Twin has represented itself as a new concept in the history of smart technologies' developments that has been thought of new solutions for

M. Poursoltan (✉) · M. K. Traore · B. Vallespir
University of Bordeaux, CNRS, IMS, UMR 5218, 33405 Talence, France
e-mail: milad.poursoltan@ims-bordeaux.fr

N. Pinède
Bordeaux Montaigne University, MICA, 33600 Pessac, France

© The Author(s), under exclusive license to Springer Nature Switzerland AG 2023
B. Archimède et al. (eds.), *Enterprise Interoperability IX*, Proceedings of the I-ESA
Conferences 10, https://doi.org/10.1007/978-3-030-90387-9_12

cyber and real-world fusion. As an emerging CPHS enabler, Digital Twin is able to depict, pause, resume, save and restore the current states of human and real objects in the cyber world, make the simulation of real-world scenarios in a cyber world and apply decisions in the real world. To benefit from these advantages, Digital Twin has an increasingly important area in CPHS. However, a major problem with the application of Digital Twin in the CPHS is the lack of a framework in order to build, conduct and develop it in the CPHS. A considerable amount of literature has tended to focus on Digital Twin as a techno-centric concept and study Digital Twin for physical assets rather than integrated systems. Some other studies structured the Digital Twin for specific aspects of the cyber-physical system (CPS) and considered humans as a side factor. This paper gives thought to Digital Twin as a cyber-human and service-centered system that benefits a bilateral data flow with human and physical parts in order to mirror and controlling of CPHS.

Embedding Digital Twin in CPHS encompasses some areas of concern. It raises several open research questions about process design, development process updating, synchronization and configuration of models, as well as the role of humans that have not yet been thoroughly explored. Due to the fact that the Digital Twin uses the meta-model concept to merge different cyber tools together in a seamless environment, the architecture-focused concept of MDA seems to be an appropriate way to decompose and develop Digital Twin. In contrast to the MDA of software development that focuses on code generating, the provided MDA zooms in on process flow and model management. In other words, it is applied to show how to map needs to the submodels at the appropriate level of automation and how to make models at appropriate level of self-synchronization and self-configuration.

This study is a part of work with the purpose of providing a methodological framework and platform for learning, validation and improvement of CPHS based on modeling.

The overall structure of the study takes the form of five sections. Section 1 provides literature review, Sect. 2 describes the position of Digital Twin in the CPHS and its subsystems, Sect. 3 illuminates the role of human in the Digital Twin, Sect. 4 is concerned with our proposal MDA, finally, the conclusion gives a brief summary and areas for further researches.

2 Literature Review

The term "cyber-physical system" has been introduced by the National Science Foundation in the United States in 2006 in order to guide a new generation of engineered systems [18, 21]. In recent years, there has been an increasing amount of literature on CPS, and subsequently, diverse forms of this concept have been developed for different application fields. (e.g., medical cyber-physical systems (MCPS), cyber-physical production systems (CPPS), etc.). CPHS has received more attention as a pivotal CPS form in various fields. There are disagreements in the literature on the term of CPHS, and it has been called by different names like human cyber-physical systems (e.g. [11, 23]), Human in the loop cyber-physical systems (e.g. [9, 28]) and human-centered cyber-physical systems (e.g., [1, 14]) however they are of the same opinion that CPHS is an arrangement of human, cyber and physical parts to perform tasks with the aim of achieving specific goals [17, 24, 33]. We would like to go one step beyond this idea and propose a more comprehensive definition for CPHS. We represent it as a system of systems (SoS) that combines humans as an integral element within cyber and physical systems and links behavioural patterns and human science in the different areas of the cyber and dynamic physical world in order to design or improve CPS with respect of the humans' diverse physical, cognitive and social capabilities. Nunes et al. [24] have presented a reference model to specify processes in order to define the human roles in the CPHS. This model includes three main processes termed: data acquisition, state interface and actuation. The first process refers to gathering data from humans, the second process addresses the processing of acquired data in order to show human physical and psychological states. Finally, actuation deals with the actions that may be performed in the system. Recently, the role of humans in the CPS has been more highlighted in manufacturing than in other fields. (e.g. [8, 10, 12, 31]). This may come from the fact that the current trend of Industry 4.0 tends to change the role of humans in manufacturing systems.

Costa et al. [5] argue that humans have always taken part in the manufacturing processes, and the integration of humans into the cyber-physical production system is a challenging task; nevertheless, the recent advances in technologies introduce several solutions to facile this integration. Krugh and Mears [17] believe that despite the undeniable role of humans in smart manufacturing systems, the human's role has not been clearly defined in CPS, so it tries to build a complementary cyber-human system to provide a unified CHS/CPS architecture for smart manufacturing. Communication within the cooperation between human and technological actors through CPS has been discussed in [2]. Lee et al. [19] have presented a 5-level CPS architecture for developing and deploying the CPS in manufacturing systems and have placed digital models at the cyber level of this architecture as a central information hub.

Digital Twin is a technology that provides high-fidelity mirroring of physical entities in cyberspace for various purposes such as simulation, real-time synchronization, virtualization, and communication [12, 13, 15, 22, 25, 26, 32]. Three subcategories of the Digital Twin have been identified in [16] based on the differentiation in terms of their levels of integration as follows: Digital Model that has no automatic data flow between physical and cyber objects, digital shadow that possesses an automatic data flow from physical to cyber objects and Digital Twin that benefits bilateral automatic data flow between physical and cyber parts.

In the last few years, some frameworks have been brought up for Digital Twin design, application and, development for various purposes. Zheng et al. [32] have proposed an application framework of Digital Twin for product lifecycle management that encompasses physical space, virtual space, and information-processing layer. They have placed bidirectional mapping, intelligent decisions, and interoperability between physical and virtual space into their framework. A unified Digital Twin framework has been proposed in [26] for the real-time monitoring and evaluation of manufacturing systems. This framework has been developed within software-defined control approach using a set of centralized data management infrastructures, a central controller, and a set of applications. Digital Twin has been placed in a central controller as the key piece of a software-defined control. A reference framework has been reported in [15] for developing the Digital Twin of physical entities, which are parts of CPS. This reference regards the high-level purpose of Digital Twin as concrete services and represents four main blocks (virtual entity platform, data management platform, physical entity platform), and service platform, to structure Digital Twin within CPS.

The challenges of Digital Twin development have been discussed in [25], and a Digital Twin structure has been developed within a CPPS. This structure defines "plant model abstraction manager" and "network component models" in order to manage and coordinate between twins, while each twin comprises models and controller.

To the best of our knowledge, there has been no structured study about the architecture of the Digital Twin in the CPHS and far too little attention has been paid to the role of humans in the Digital Twin. To cover these issues, a new architecture using the MDA approach is developed in the current study to take humans into account in order to design and develop the Digital Twin for CPHS.

3   Digital Twin in the CPHS

Multiple, heterogeneous, distributed,and occasionally independently operating systems that are embedded in a network, form of SoS [6]. As a type of SoS, the CPHS consists of humans, cyber and physical systems, and the relationship between these systems are established through three subsystems termed cyber-physical, cyber-human and human-physical system [33]. According to [27], physical systems refer to the natural and human-made systems built through the laws of physics and operating in continuous time. These systems generally are composed of physical objects, sensors, actuators and communication networks. Cyber systems are related to computational systems carried out in cyberspace. As a cyber component, Digital Twin is placed in the cyber space [19] and is able to interact bilaterally with human and physical components with the help of cyber-physical and cyber human systems, as shown in Fig. 1.

Cyber systems should proactively use the help of humans to perform the operations when needed [29]. The cyber-human system (CHS) aims to acquire data from humans and feedback information to humans in order to enhance control and monitoring of CPS or assist humans in performing their jobs more safely and efficiently in the real-world [17]. Human-physical systems (HPS) are formed when human needs to be equipped with the physical systems to communicate with cyber systems (e.g., when an individual carries a GPS tracker) or human's task must be performed with the help of physical systems (e.g., using the elevator) or physical systems requires human to

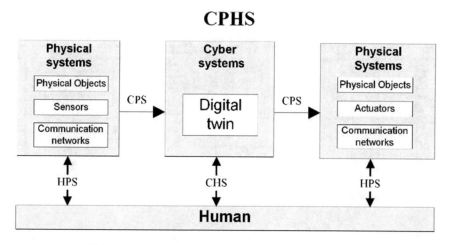

Fig. 1 Position of Digital Twin in the CPHS

complete jobs (e.g. semi-automated robotic systems). Like CHS, the CPS has a two-way relationship between cyber and physical systems for receiving and sending data. Figure 1 illustrates CPHS components and their subsystems.

Every system with a large network of components, many-to-many communication channels, and sophisticated information processing, can be referred to as complex which makes the prediction of system states difficult [20]. Adopting this definition, we can say that Digital Twin is a complex system since it deals with heterogeneous and massive amounts of data and various data processing, multiple models and applications, as well as diverse communication links with physical systems and humans. Humans get involved in many cyber systems by design or implication [7]. This indicates a need to understand the various perceptions of humans that exist for the design and operation of the Digital Twin. Participation of human in Digital Twin can occur through design, computation, communication, and control. To represent and organize our knowledge about human's roles in Digital Twin, we use graphical concept mapping, which comprises concepts (represented by boxes) and relationships between pairs of concepts (represented by labeled links) (see Fig. 2).

Generally, models in Digital Twin are used to communicate an understandable situation of the real world to perform analysis and resolve problems as well as keep knowledge to be used in a current or further situation. Each model differs in some form of functionality, complexity, integrations and, technologies. The goals of models in Digital Twin can be categorized into optimization, diagnostic, and prognostic. It is essential to evaluate how well cyber systems mirror their physical counterparts. One of the most critical processes that human gets involved in is to verify and validate the models in Digital Twin. Given that humans do not perform a task in the same manner and also because of human performance constraints and cognitive limitations, the presence of humans in such a complex system makes it difficult to predict the behaviour of the system. However, Digital Twin needs to be evolved so it should

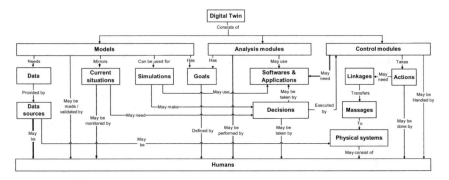

Fig. 2 Digital Twin concept map for CPHS

be flexible and human adaptability can also play a significant role in Digital Twin's resilient system and also keep the CPSH stable.

4 Proposed Framework

The proposed methodological approach for this study is a mixed methodology based on MDA and SD logic. The using of both concepts brings organized model development as well as interactive value creation for CPHS.

The central idea behind an MDA is to separate the specification of system from the platform details that system uses to be executed [4]. Platform is defined as the methods, technologies and subsystems under which system is lunched. Chong et al. [3] introduce reusability, interoperability and portability as three principle goals of MDA through architectural separation. Tyson et al. [30] have used three categories of interoperability for service and data interoperability as technical interoperability, semantic interoperability and process interoperability. They believe that model-driven conceptual architecture can be utilized to deal with interoperability challenges and provides better integration of SoS and also facilitates communication between different components of a SoS.

A Digital Twin is not deployed on a unique platform, so it should be designed and developed from a platform-independent perspective. This reveals that Digital Twin is placed at the level of the platform-independent model (PIM) of MDA. The implementation of such an architecture is performed through heterogonous platform-specific models (PSMs). Thus, developing Digital Twin through the MDA approach provides not only an early evaluation of the overall system and focuses on platform independent solutions before full implementation but also provides interoperable capabilities for its platform dependent models in order to be developed under their specific platform.

In addition, the applied MDA approach allows the Digital Twin structure to be generic enough to do analysis without worrying about technologies and cyber systems in which models and applications will be executed.

The high-level Digital Twin's target is to provide concrete services [15]. The foundational proposition of SD logic for Digital Twin is that cyber and real space are fundamentally concerned with the exchange of service. The key to these services is to create value through mutual cooperation. In other words, human, cyber and physical parts cannot deliver individually the value of Digital Twin to the CPHS but can participate in the creation and offering of value propositions. The successful implementation of Digital Twin will need to create value and require the right mix of data, models and actions.

Provided architecture (see Fig. 3) includes three phases termed feeding, modelling, and servicing comprising data, model and action modules. Each module has a process and actor dimensions. The overall level of participation of cyber actors indicates the degree of automation of the system. The data managing process deals with activities regarding access, storing, updating, and ensuring data reliability. Modelling has various processes to ensure appropriate mirroring and analyzing situations. Each

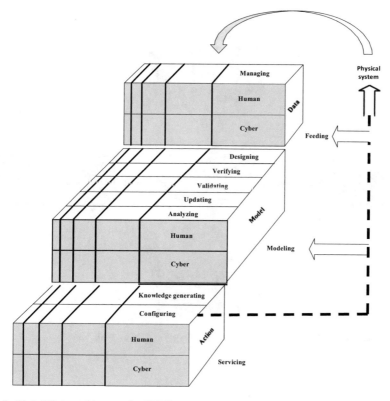

Fig. 3 Digital Twin architecture for CPHS

model in Digital Twin may serve differently so the activities of each process may vary from one model to another. Final services may lead to knowledge generation or reconfiguration of cyber or physical systems. Knowledge generation is related to the result of modelling which do not bring any configuration into systems (e.g., when scenario testing doesn't show desirable result), but the results need to be conserved as knowledge.

Reconfiguration may occur in the cyber or physical systems. Any reconfiguration in physical systems may change the behaviour of physical systems and, consequently, may impact the behaviour of cyber systems.

The usage of the framework can be illustrated briefly by an example (see Fig. 4). Patient, ambulance, nurse and hospital equipped with sensors or/and communication equipment are the sources of data that feed the cloud database. Data is processed in cloud environments. By using real-time data from the cloud, the practitioner runs the digital models of CPHS's components. Analysis may be performed to diagnose abnormalities, predict the future condition of the patient or estimate medical care arrival time. Decisions may lead to adjust electrics patient devices, coordinating services in the hospital, sharing the patient's condition with the hospital, alerting the ambulance and providing recommendations for the nurse.

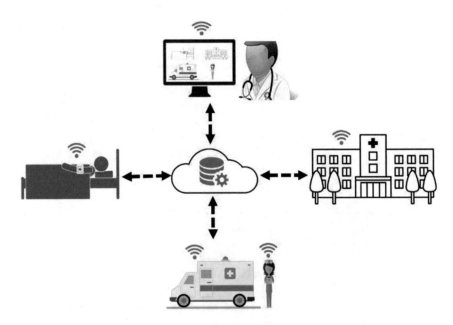

Fig. 4 Illustration of the use of the Digital Twin architecture in a CPHS

5 Conclusion

As pointed out in the first of this paper, the aim was to answer the question of in what way Digital Twin should be designed and developed for a CPHS. This paper has argued the position of the Digital Twin in the CPHS and provided a new Digital Twin concept map. Then we proposed a new framework and illustrated how value could be generated through services in Digital Twin in a cooperative way. The ideas discussed in this article reflect the thinking of humans in the loop. If the debate is to be moved forward, a better understanding of how Digital Twin can be optimized by human. Considerably, more work will need to be done to determine the level of automation of Digital Twin in the CPHS. In addition, it would be interesting to assess the effects future research will explore the role of Digital Twin in a value-producing system within the context of decentralization and autonomy, where humans actor plays a key role.

References

1. Ansari, F., Khobreh, M., Seidenberg, U., & Sihn, W. (2018). A problem-solving ontology for human-centered cyber physical production systems. *CIRP Journal of Manufacturing Science and Technology, 22,* 91–106.
2. Böhle, F., & Huchler, N. (2017). *Cyber-physical systems and human action: A re-definition of distributed agency between humans and technology.* Elsevier.
3. Chong, S., Wong, C. B., Jia, H., Pan, H., Moore, P., Kalawsky, R., & O'Brien, J. (2011, August). Model driven system engineering for vehicle system utilizing model driven architecture approach and hardware-in-the-loop simulation. In *2011 IEEE International Conference on Mechatronics and Automation* (pp. 1451–1456). IEEE
4. Cloutier, R. (2008). Model driven architecture for systems engineering. In Presentation Slides, *Stevens Institute of Technology, presented at INCOSE International Workshop.*
5. Costa, D., Pires, F., Rodrigues, N., Barbosa, J., Igrejas, G., & Leitão, P. (2019). Empowering humans in a cyber-physical production system: Human-in-the-loop perspective. In: *2019 IEEE International Conference on Industrial Cyber Physical Systems (ICPS)* (pp. 139–144). IEEE.
6. Delaurentis, D. (2007). Role of humans in complexity of a system-of-systems. In: *International Conference on Digital Human Modeling* (pp. 363–371). Springer.
7. Eskins, D. & Sanders, W. H. (2011). The multiple-asymmetric-utility system model: A framework for modeling cyber-human systems. In *2011 Eighth International Conference on Quantitative Evaluation of Systems* (pp. 233–242). IEEE.
8. Fantini, P., Tavola, G., Taisch, M., Barbosa, J., Leitão, P., Liu, Y., Sayed, M. S. & Lohse, N. (2016). Exploring the integration of the human as a flexibility factor in CPS enabled manufacturing environments: Methodology and results. In *IECON 2016-42nd Annual Conference of the IEEE Industrial Electronics Society* (pp. 5711–5716). IEEE.
9. Feng, S., Quivira, F. & Schirner, G. (2016). Framework for rapid development of embedded human-in-the-loop cyber-physical systems. In *2016 IEEE 16th International Conference on Bioinformatics and Bioengineering (BIBE)* (pp. 208–215). IEEE.
10. Gaham, M., Bouzouia, B. & Achour, N. (2015). Human-in-the-loop cyber-physical production systems control (HiLCP$_2$sC): A multi-objective interactive framework proposal. In *Service orientation in holonic and multi-agent manufacturing.* Springer.

11. Gelenbe, E., Gorbil, G. & Wu, F.-J. (2012). Emergency cyber-physical-human systems. In *2012 21st International Conference on Computer Communications and Networks (ICCCN)* (pp. 1–7). IEEE.
12. Graessler, I. & Pöhler, A. (2017). Integration of a digital twin as human representation in a scheduling procedure of a cyber-physical production system. In *2017 IEEE International Conference on Industrial Engineering and Engineering Management (IEEM)* (pp. 289–293). IEEE.
13. Haag, S., & Anderl, R. (2018). Digital twin–Proof of concept. *Manufacturing Letters, 15*, 64–66.
14. Hadorn, B., Courant, M. & Hirsbrunner, B. (2016). Towards human-centered cyber-physical systems. Université de Fribourg.
15. Josifovska, K., Yigitbas, E. & Engels, G. (2019). Reference framework for digital twins within cyber-physical systems. In: *Proceedings of the 5th International Workshop on Software Engineering for Smart Cyber-Physical Systems* (pp. 25–31). IEEE Press.
16. Kritzinger, W., Karner, M., Traar, G., Henjes, J., & Sihn, W. (2018). Digital Twin in manufacturing: A categorical literature review and classification. *IFACPapersOnLine, 51*, 1016–1022.
17. Krugh, M., & Mears, L. (2018). A complementary cyber-human systems framework for Industry 4.0 cyber-physical systems. *Manufacturing Letters, 15*, 89–92.
18. Lee, E. A. (2015). The past, present and future of cyber-physical systems: A focus on models. *Sensors, 15*, 4837–4869.
19. Lee, J., Bagheri, B., & Kao, H.-A. (2015). A cyber-physical systems architecture for industry 4.0-based manufacturing systems. *Manufacturing letters, 3*, 18–23.
20. Mitchell, M. (2009). *Complexity: A guided tour*. Oxford University Press.
21. Monostori, L. (2018). Cyber-physical systems. In: S. Chatti & T. Tolio (Eds.), *CIRP encyclopedia of production engineering*. Springer Berlin Heidelberg.
22. Negri, E., Fumagalli, L., & Macchi, M. (2017). A review of the roles of digital twin in cps-based production systems. *Procedia Manufacturing, 11*, 939–948.
23. Netto, M., & Spurgeon, S. K. (2017). Special section on cyber-physical and human systems (CPHS). *Annual Reviews in Control, 44*, 249–251.
24. Nunes, D., Silva, J. S. & Boavida, F. (2018). *A practical introduction to human- in-the-loop cyber-physical systems*. Wiley Online Library
25. Park, H., Easwaran, A., & Andalam, S. (2018). *Challenges in digital twin development for cyber-physical production systems*. Springer.
26. Qamsane, Y., Chen, C.-Y., Balta, E. C., Kao, B.-C., Mohan, S., Moyne, J., Tilbury, D. & Barton, K. (2019). A Unified digital twin framework for real-time monitoring and evaluation of smart manufacturing systems. In *2019 IEEE 15th International Conference on Automation Science and Engineering (CASE)* (pp. 1394–1401), IEEE.
27. Sacala, I. S., Moisescu, M. A., Munteanu, I. D. C. A., & Caramihai, S. I. (2015). Cyber physical systems-oriented robot development platform. *Procedia Computer Science, 65*, 203–209.
28. Schirner, G., Erdogmus, D., Chowdhury, K. & Padir, T. (2013). The future of human-in-the-loop cyber-physical systems. *Computer* 36–45.
29. Sukkerd, R., Garlan, D. & Simmons, R. (2015). Task planning of cyber-human systems. *SEFM 2015 Collocated Workshops* (pp. 293–309), Springer.
30. Tyson, G., Taweel, A., Zschaler, S., Staa, T. V. & Delaney, B. (2011). A model-driven approach to interoperability and integration in systems of systems. In *Proceedings of Work- shop on Model-Based Software and Data Integration (MBSDI)*.
31. Wang, L., Törngren, M., & Onori, M. (2015). Current status and advancement of cyber-physical systems in manufacturing. *Journal of Manufacturing Systems, 37*, 517–527.
32. Zheng, Y., Yang, S., & Cheng, H. (2019). An application framework of digital twin and its case study. *Journal of Ambient Intelligence and Humanized Computing, 10*, 1141–1153.
33. Zhou, J., Zhou, Y., Wang, B., & Zang, J. (2019). Human–cyber–physical systems (HCPSs) in the context of new-generation intelligent manufacturing. *Engineering, 5*, 624–636.

Interoperability of Processes

Interoperability Concerns for Multidimensional Urban Mobility Within the Frame of MaaS

Faheem Ahmed Abassi, Hedi Karray, Raymond Houe, Muhammad Ali Memon, and Bernard Archimède

Abstract Nowadays, due to urbanization growth, the need for mobility arises around the world. Some cities indeed are seeking for innovative solutions in order to meet the increasing users demand in connectivity, among which mega cities that have introduce air mobility. This latter will increase the mobility externalities and complexify its management. In the past decade, mobility as a service paradigm has been proven as the best approach to address such issues. But the current solutions are provided by autonomous mobility providers. In order to provide policy-makers in cities with a decision support tool allowing them to manage traffic regulation, environmental pollution, safety of the passengers, and services and infrastructures renewal, there is a need to address interoperability issue between the existing mobility systems. This paper is a preliminary study of interoperability concerns in the context of multidimensional urban mobility, which includes land and air modes. To that end, we present and discuss the building blocks of the underlying system and show which kinds of the interoperability occur and provide directions to solve them, within the frame of mobility as a service (MaaS).

Keywords Mobility as a service · Urban air mobility · Multidimensional urban mobility · Interoperability

F. A. Abassi (✉) · H. Karray · R. Houe · B. Archimède
University of Toulouse, INP-ENIT, 47 Avenue d'Azereix, BP1629, 65016 Tarbes Cedex, France
e-mail: faheem.ahmed@enit.fr

H. Karray
e-mail: mkarray@enit.fr

R. Houe
e-mail: raymond.houengouna@enit.fr

B. Archimède
e-mail: bernard.archimede@enit.fr

F. A. Abassi · M. A. Memon
Institute of Information and Communication Technology, University of Sindh, Sindh, Pakistan
e-mail: muhammad.ali@usindh.edu.pk

© The Author(s), under exclusive license to Springer Nature Switzerland AG 2023
B. Archimède et al. (eds.), *Enterprise Interoperability IX*, Proceedings of the I-ESA
Conferences 10, https://doi.org/10.1007/978-3-030-90387-9_13

1   Introduction

In recent years, we have witnessed a rapid growth of urbanization which poses a major challenge to the cities. An alarming increase in population, economic, social, environmental, and traffic-related problems is becoming more acute across the globe, particularly in mega cities. These latter were looking for a solution to ever-growing needs of the population, developed with hopes of economies of scale, both for the governments and the businesses (construction, manufacturers, suppliers, etc.) that helped build them. Another advantage was vicinity of the public to great resources and occupations that these cities would provide. Industries were welcomed to these cities since they knew the cities could hold their workforce and provide a flourishing environment both for the workers and the enterprises.

The mounting pace of urbanization threatens infrastructure of cities (e.g., it renders transportation system inadequate and ineffective). This creates a number of other problems which prove significantly harmful to the lives of people and to their financial stability. Owing to the fact that the need of urban mobility arises for urban planning decision-makers to solve transportation challenges, such as traffic congestion, safety of the passengers, environmental pollution, and infrastructures renewal. Urban mobility states to the effective movement of people and goods, by well-organized, environmentally good, safe, and reasonable transportation that contributes to improving social fairness, public health, resilience of cities, and efficiency. Two-dimensional transportation and mobility are recognized as central to sustainable development since they increase economic progress, enhance accessibility, and achieve better integration of the economy while regarding the situation. Better transport encourages universal access to social services and therefore can make an important contribution to merging and achieving development advantages in urban areas. In the foreseeable future, decision-makers will introduce an efficient deployment of the new mobility paradigm, which includes the air mode. Urban air mobility (UAM) refers to as a third dimension of the mobility which is a significantly effective solution for the problems of areas where merely increasing two-dimensional capacity cannot tackle enduring traffic problems. It also creates new opportunities for travelers for whom personal comfort and speed are at a premium, as well as for rescue services. Moreover, gradual merger of urban air mobility with existing mobility landscape would pave the path for smooth and safe travel. It would give the passenger tremendous experience at an increasingly low cost [1]. As a matter of fact, there are various private mobility providers who deliver large number of mobility services in the context of multidimensional urban mobility (MUM). In order to facilitate public authorities to enhance services and infrastructures availability and quality, for the future mobility, the systems managing the current mobility services need to be interoperable, so as to manage the underlying complexity.

The rest of the paper is structured as follows: Sect. 2 highlights the related work associated with urban mobility concerns, including a discussion on how MaaS addresses such concerns, while Sect. 3 presents some interoperability concerns at different levels for MaaS. Finally, Sect. 4 concludes the paper and provides directions for future work.

2　Related Work

This section outlines the main studies related to MUM and the associated interoperability issues.

As stated earlier, the rapid growth of urbanization spawns a variety of MUM concerns which must be tackled timely, such as traffic congestion, infrastructure, safety of the passengers, and environmental pollution (as shown in Fig. 1).

- **Traffic congestion**. Urbanization creates a lot of traffic-related problems. Congestion has proved to be a significant issue. Surely, if there is a constant increase in the number of vehicles while the road system and parking areas remain the same, commute will become difficult. This particularly occurs in the urban areas; however, the problem of congestion cannot be eradicated by merely initiating infrastructure projects, such as bridges, roads, and railway networks. Technology has evolved as a tool to solve human problems and making lives easier, and it is greatly helpful in reducing congestion as well. It is very important to note other factors that are responsible for traffic jams like accidents, maintenance work, ineffective transportation systems, etc., which needs to be tackled as well, both individually and a part of the overall solution [2].
- **Infrastructures**. There is a chain of railway networks, roads, footpaths, airports, and other infrastructure projects which facilitate transportation. These need to be planned according to the projected needs of each community and in places where new projects are not possible, rejuvenation or at the very least up-keep projects, and it can be implemented to make sure that the systems work as expected, helping prepare for increased or decreased flow accordingly [3].

Fig. 1　MUM concerns

- **Safety of the passengers**. Many accidents take place due to increase in traffic. Often, the more congested the traffic in urban areas is, the more the accidents, injuries, and deaths are probable. Some countries have experimented with rules like allowing only a specific segment of cars (e.g., cars with odd or even registration numbers) on roads on certain days, but people bypass these laws by buying multiple cars. There is also a decreased sense of security among the commuters [4].
- **Environmental pollution**. Energy consumption has colossally enhanced due to urban transportation. Therefore, pollution has increased. Coupled with vexing noise, pollution has rendered life of urban people miserable as it is gravely injurious to their health [5].

The aim of city planners is to improve cities' management of natural and municipal resources and in turn the quality of life of their citizens. A city that performs well in the economy, people, governance, mobility, environment, and living, and is built on a clever combination of endowments and activities of self-decisive, independent, and aware citizens [6]. Finding a way to deal with above cited MUM concerns, city planners need some smart urban mobility solutions such as MaaS.

2.1 Mobility as a Service

In the vision of city, MaaS is globally a new way of structuring urban mobility that meets sustainability requirements, since its intended purpose is to prevent individuals from using their own vehicles. As a counterpart, a wide range of services is offered to them. Indeed, it is based on a wide use of digital technologies to guarantee access to information for users and the invoicing of the services used, within the frame of sustainable development. It enables the users to easily find the best route, price, multimodal framework across several end-to-end services (through convenient tools such as recommender systems or routing planners), and real-time information such as traffic condition time of day and demand. MaaS also organizes the relationships of the urban mobility stakeholders and ensures that their respective priorities are met: (i) the end user prioritizes speed and cost for his or her travel, without sacrificing comfort and reliability; (ii) the transport authority must ensure accessibility to the city's various attractions while reducing costs to make the best use of public funds, in a context where environmental concerns and their impact on health (pollution) and climate (carbon footprint) are becoming increasingly important; (iii) mobility operators, public and/or private, highlight the need for profitability of the services they offer in order to be able to invest and pursue the development of services that are increasingly in line with users' expectations [7]. Righteous cycle in relation to these stakeholders' priorities can then occur: (i) services provided to end users motivate them leaving their private cars for public modes; (ii) the city center is then relieved of congestion, which also reduces the carbon footprint, while (iii) users switch to high-performance services whose costs are controlled, taking into consideration this mass modal shift; (iv) operators in turn can continue their development and investments.

According to Jittrapirom et al. [8], the following features can characterize MaaS:

- **Integration of transport modes**. The objective is to encourage the use of public transport services through multimodal transport and to facilitate intermodal travel;
- **Tariff option**. It is composed of two types of fare, "mobility package" and "pay-as-you-go"; the first contains packages of different modes of transport and includes a number of km/minutes/points that can be used in exchange for a monthly payment, while the second charges users according to their actual use of the service;
- **Single platform**. It is based on digitalization of content through which users can easily access (including from their smartphone) various services such as travel planning, booking, ticketing, e-payment, and real-time traffic information;
- **Multiple actors**. Interaction occurs through a digital platform between different stakeholders within MaaS ecosystem, including individual or corporate customers, transport service providers (private or public), platform owners (third parties, public transport providers, or metropolitan authorities), e-payment, e-ticketing, telecommunications, and data management companies;
- **Use of Internet technologies**. It is mainly based on the combination of devices such as smartphones or computers, mobile Internet network (Wi-Fi, 4G, LTE, GPS) e-ticketing and e-payment systems, database management systems, and infrastructure integrated technology (Internet of Things);
- **Obligation to register**. The aim is to enable the end user to join the platform and benefit from access to services, including personalized services;
- **On-demand services**. The purpose is to facilitate the satisfaction of end user requirements and expectations;
- **Personalization**. This allows end users to change service options according to their preferences, hence the possibility of freely composing related trips or building their mobility package, with a different volume of use for certain modes of transport.

It is clear that MaaS is mainly infused by Internet and Communication Technologies (ICT), used in both backward and forward applications, to support the optimization of traffic fluxes, but also to gather citizens' views about livability in cities or quality of local public transport facilities [9]. By using these new technologies, the need arises to introduce new type of services such as car sharing, bike sharing, and ride sharing [8]. Several MaaS-based platforms have been developed: for example, BeMobility at Berlin, EMMA at Montpellier, OptiMod at Lyon, STIB at Brussels, SHIFT at Las Vegas, SMILE at Vienna, UbiGo at Gothenburg, etc. [2, 8, 10]. Although most of them have succeeded in implementing the integration of e-ticketing, e-payment, several mobility modes, and the development of practical solutions for users such as itinerary recommendations, proposal of multimodal solutions, and real-time traffic information, these achievements did not address the complexity induced by the third mobility dimension (i.e., air mode), nor its management, which is required for decision-makers in cities. Within the frame of sustainable development,

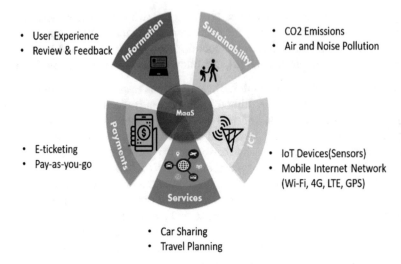

- User Experience
- Review & Feedback

- CO2 Emissions
- Air and Noise Pollution

- E-ticketing
- Pay-as-you-go

- IoT Devices(Sensors)
- Mobile Internet Network (Wi-Fi, 4G, LTE, GPS)

- Car Sharing
- Travel Planning

Fig. 2 MaaS ecosystem

air mobility will undoubtedly induce the need for shared poles of exchange between land and air modes, and also the necessity to rethink the pricing schemes (even the underlying business model), which may complexify the overall management of services along with the infrastructure maintenance.

It appears then that five key characteristics should keep in mind to frame a MaaS (as shown in Fig. 2).

- **Information**. To meet users' needs on the basis of in-depth information on the reasoning behind their views and to explain their experiences using the service more systematically. Inspiring participations such as giving away free bus passes [11], often combined with information to increase the success rate, is an example of convenient information service provided.
- **ICT**. ICTs have played a vital role in the transformation process from old technologies to new trends of technologies such as IoT devices, e.g., sensors and actuators to collect real-time data for MaaS providers.
- **Sustainability**. The aim of MaaS is to sensitizing users to relieve their personal cars; carbon footprint reduction be the consequence to provide sustainable environmental model, to manage environmental issues such as air and noise pollution.
- **Smart Services**. MaaS platform should provide several smart services by using single platform such as car sharing, bike sharing, best route suggestion, mix-modal transportation, travel planner, and e-ticketing.
- **Payments**. Pricing and payment would be available in more convenient and an efficient manner, for instance, to provide a single price for the same ticket, and that the payment is digitalized.

The aim of the present work is therefore to develop a decision support system allowing to help policy-makers in cities to efficiently manage the above-mentioned issues, taking into consideration models provided by the existing mobility solutions. This requires to cope with several levels of interoperability within the intended support system.

In an implementation perspective, the above-mentioned MaaS characteristics can finally be seen as different components of the underlying system.

2.2 Brief Overview of Interoperability Concerns and Approaches

In order to make MaaS framework components interoperable, it is necessary to consider different interoperability concerns, i.e., data, services, process, and business and different interoperability implementation approaches, i.e., integrated, unified, and federate [12]:

- **The data interoperability**. Generally, consider the main interoperability concern related to data access, aggregation, and reasoning. It is about to find and share information coming from cross-domain sources, i.e., databases, operating systems, and database management systems.
- **Interoperability of services**. It is referred as services that are independently developed by different vendors and running together to solve syntactic and semantic level issues.
- **Interoperability of processes**. It is referred as a combination of the different services that work together. Process defines in which order services will be running according to user needs. Mostly, several processes are functioning collaboratively within an organization to validate certain tasks.
- **Interoperability of business**. The interoperability of business refers to the workflows of the system in a consistent way for business-to-business integration.

Following are approaches to address above concerns.

- **Integrated approach**. Implementing interoperability over an integrated approach means that different models used same template. The common format is not necessarily an international standard but must be agreed by all stakeholders to develop models and build systems.
- **Unified approach**. It means there is a shared format between systems, but it only exists at high level (abstraction). This format is not an executable likewise in integrated approach.
- **Federated approach**. In this approach, there is no shared format between all the systems, to make systems interoperable at run-time. It means federated approach suggests that no partner enforces their models, languages, and methods of work, and they must share an ontology with each other [12].

There are several ontologies in literature dealing with several interoperability concerns in the domain of mobility by using different approaches as mentioned above that includes traffic management, accidents on roads, and transport problems, etc. [13].

The Ontology for Transportation Networks (OTN) was introduced [14] as part of the reasoning on the web with rules and semantics (REWERSE) project. OTN formalizes and extends the Geographic Data Files (GDF) for geographic information and addressing data and service level concerns using integrated approach. The Transport Disruption Ontology is calculated to accumulate data and help merge it so as to identify events which can create disruption in traveling. This ontology was used in Social Journeys in order to unearth in what way social media could be helpful for sharing information to the commuters and only focused data level concerns using integrated approach [15]. Ontology-based management of the traffic on roads was established to help drivers take proper decisions, with the ultimate objective of making the way effectively clear for emergency vehicles and resolve, data and services level concerns by using unified approach [16]. Osmonto ontology is used for OpenStreetMap tags and trying to solve location-based service interoperability concern using unified approach [17]. GenCLOn was built and presented as an ontology that dealt with city logistics. GenCLOn is designed to encourage the sharing and reutilization of the paradigms constructed to guess the behavior of all parties that participate in the area of urban logistics using federated approach to solve data and business level interoperability concerns. Recent work by Benvenuti et al. [18] merges KPIOnto and Trans-model ontologies to strengthen monitoring of system of public transportation. KPIOnto catches generic concepts connected to Key Performance Indicators (KPIs). Trans-model and KPIOnto are connected by linking the basic data classes in Trans-model with indicators from KPIOnto. KPIOnto and Trans-model are part of a suggested frame for a system to buttress the design and dissection of a management system for the systems of public transportation. This work also solves data level interoperability issues using integrated approach.

Above all cited ontologies in literature used different approaches to solve different level of interoperability concerns. In addition, the aim of this research work is to discuss different interoperability concerns for MUM within the frame of MaaS.

3   Interoperability Concerns for MUM Within the Frame of MaaS

In this section, we discussed different building blocks of the MaaS ecosystem and different interoperability concerns associated with these blocks. MaaS building blocks comprise stakeholders, operating infrastructure, and smart services.

- **Stakeholders**. Potential stakeholders of MaaS are users, providers, and public authorities (as shown in Fig. 3). End user used different services that are provided by various mobility providers and public authorities such as government need to

make new policies and regulation to address MUM concerns like traffic congestion, infrastructure, safety, and pollution to creating a sustainable green and
user-friendly environment [9]. From a stakeholder point of view, they need to
exchange and share data by using some services. For instance, public authorities need services data from different mobility providers linked to existing city
infrastructure to make policies and regulation to improve quality of existing infrastructures. From a mobility provider perspective, they need to integrate booking
and payment processing systems that are built separately by different solution
providers. Mobility providers are always looking for an opportunity to generate
new business models from the existing model and add a new business model on
top of the existing model. For example, the pay-as-you-go model is a new way of
payment and it must be interoperable with traditional payment systems.

- **Operating infrastructure**. Contains physical objects such as IoT devices, e.g.,
  sensors and actuators, infrastructure of roads, bridges, railway stations, airports,
  and vehicles network, i.e., 1D, 2D and 3D, etc. (as shown in Fig. 4), are utilized to
  analyze the environment in order to collect information with the help of sensors
  and initiate actions so as to impact the environment and give a response back
  to systems [19]. In operating infrastructures, we have different IoT devices,
  i.e., sensors and actuators to collect data from various heterogeneous sources
  to make them interoperable for analysis and take some appropriate decisions, i.e.,
  infrastructure renewal.
- **Smart services**. The new mobility paradigm is changing the core mobility services
  like public transport, car rental, parking, taxis, and shuttle into new smart services
  such as car sharing, bike sharing, integrated mobility, and on-demand mobility
  (as shown in Fig. 5) [20]. Since these all services are developed by autonomous

Fig. 3 Stakeholders

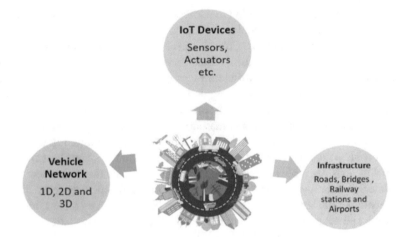

Fig. 4 Operating infrastructure

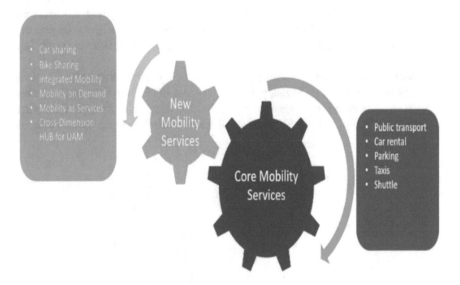

Fig. 5 Smart services

solution providers, to use all these services on a single platform, need to be integrated and interoperable with each other, the user of the services needs to adopt the different processes to use a particular service like first he registers then pay. So, these processes are also interoperable for smooth accessing of the service.

Thus, there is a need to provide standards to make blocks interoperable with each other to ensure that the interfaces, data flows, and message content allow for the open exchange of communications and collaboration among MaaS building blocks.

4   Conclusion and Future Perspectives

This paper has presented a conceptualization of different interoperability concerns such as data, services, business, and organization for MaaS building blocks in the context of MUM. The future perspective of this research is to federate ontology based on available ontologies in literature using semantic model. Build intelligent decision support system for decision-makers to manage traffic regulation, environmental pollution, and safety of the passengers and infrastructures renewal.

References

1. Baur, S., et al. (2018). Urban air mobility the rise of a new mode of transportation, November.
2. Litman, T. (2016). Smart congestion relief comprehensive evaluation of traffic congestion costs, February.
3. Bryceson, D. F., et al. (2003). Sustainable livelihoods, mobility and access needs. *Department of International Development* (July), 57.
4. Sawin, J., Martinot, E., & Appleyard, D. (2010). Global status report. *Renewable Energy World, 13*(5), 24–31.
5. Safonov, P., Favrel, V., & Hecq, W. (2002). Environmental impacts of mobility and urban development: A case study of the Brussels-capital region. *Managing for Healthy Ecosystems* 741–755.
6. Khatoun, B. Y. R., & Zeadally, S. (2016). Cities.
7. Hietanen, S. (2014). 'Mobility as a service'—The new transport model? *Eurotransport, 12*(2), 2–4.
8. Jittrapirom, P., et al. (2017). Mobility as a service: A critical review of definitions, assessments of schemes, and key challenges. *Urban Planning, 2*(2), 13.
9. Benevolo, C., Dameri, R. P., & D'Auria, B. (2016) Abstract. Smart mobility in smart city action taxonomy, ICT intensity and public benefits. *Adhyayan: A Journal of Management Sciences, 4*(1), 13–29.
10. Kamargianni, M., Li, W., Matyas, M., & Schäfer, A. (2016). A critical review of new mobility services for urban transport. *Transportation Research Procedia, 14*, 3294–3303.
11. Fujii, S., & Kitamura, R. (2003). What does a one-month free bus ticket do to habitual drivers? An experimental analysis of habit and attitude change. *Transportation, 30*(1), 81–95.
12. Chen, D. (2017). Framework for enterprise interoperability. *Enterprise Interoperability: INTEROP-PGSO Vision* 1–18.
13. Katsumi, M., & Fox, M. (2018). Ontologies for transportation research: A survey. *Transportation Research Part C: Emerging Technologies, 89*(September 2017), 53–82.
14. Lorenz, B. (2005). A1-D4 ontology of transportation networks.
15. Corsar, D., Markovic, M., Edwards, P., & Nelson, J. D. (2015). The transportdisruption ontology. In *Lecture Notes in Computer Science (including subseries Lecture Notes in Artificial Intelligence and Lecture Notes in Bioinformatics)* (Vol. 9367, pp. 329–336).
16. Bermejo, A. J., Villadangos, J., Astrain, J. J., & Córdoba, A. (2013). Cit. *Studies in Computational Intelligence, 446*, 103–108.
17. Kutz, O., Codescu, M., Vale, D. C., & Mossakowski, T. (2014). OSMonto—An ontology of OpenStreetMap tags ontology-based route planning for OpenStreetMap. In *The Terra Cognita Workshop on Foundations, Technologies and Applications of the Geospatial Web, in conjunction with the 11th International Semantic Web Conference*, September.
18. Benvenuti, F., Diamantini, C., Potena, D., & Storti, E. (2017). An ontology-based framework to support performance monitoring in public transport systems. *Transportation Research Part C: Emerging Technologies, 81*, 188–208.

19. Fukuda, A., et al. (2016). Towards sustainable information infrastructure platform fors-mart mobility—Project overview. In *Proceedings—2016 5th IIAI International Congress on Advanced Applied Informatics, IIAI-AAI 2016* (pp. 211–214).
20. Bellini, F., Dulskaia, I., Savastano, M., & D'Ascenzo, F. (2019). Business models innovation for sustainable urban mobility in small and medium-sized European cities. *Management and Marketing, 14*(3), 266–277.

Empowering Process Quality Through Microservices. A ZDMP Perspective

Víctor Anaya, Francisco Fraile, Raúl Poler, and Ángel Ortiz

Abstract Machine learning is omnipresent in today's software solutions. One of the areas of interest that benefits from smart data exploitation is the manufacturing of products with zero defects. But manufacturing a product is the result of entangled processes spanning different companies that exchange products usually in not exclusive contracts. The machine learning promise is sustained by information. The challenges are clear: sharing an increasing amount of information along the supply chain while keeping competitive knowledge in house, reducing the complexity of implanting AI solutions and respecting heterogeneous-distributed diverse existing software systems. This paper's purpose is to present an upcoming solution from the perspective that the authors are bringing to the H2020 Zero Defects Manufacturing Platform European project.

Keywords Machine learning · Zero defects manufacturing · Software architecture · Microservices · Process quality assurance · AI as a service

1 Introduction

1.1 Zero Defects Manufacturing and ZDMP

In the last years, many industrial production entities in Europe have started strategic work towards a digital transformation into the Fourth Industrial Revolution termed Industry 4.0. Based on this new paradigm, companies must embrace a new technological infrastructure, which should be easy to implement for their business and easy to implement with other businesses across all their machines, equipment and systems.

To remain competitive and keep its leading manufacturing position, European industry is required to produce high-quality products at a low cost, in the most efficient way [1]. Today, the manufacturing industry is undergoing a substantial

V. Anaya (✉) · F. Fraile · R. Poler · Á. Ortiz
Universitat Politecnica de Valencia, 46022 Valencia, Spain
e-mail: vanaya@cigip.upv.es

transformation due to the proliferation of new digital and ICT solutions, which are applied along the production process chain and are helping to make production more efficient, as in the case of smart factories. One of those areas is the zero defect manufacturing [2] where Industry 4.0 technology is applied with the purpose to detect, predict and prevent quality defects on the manufacturing process.

The purpose of the current article is to present the Process Quality Services provided as part of the ZDMP—Zero Defects Manufacturing Platform—EU project [3] and more specifically to introduce a serverless microservice architecture of process quality services based on machine learning and optimisers models with the purpose to address process quality assurance initiatives.

The presented work is a component part of the Zero Defects Manufacturing Platform, and as such is one of the microservices that a Zero Defects Application developer will use when providing specific defect avoidance apps.

ZDMP as a whole will be of value for an ecosystem, where software developers and integrators will provide solutions that will benefit from manufacturing infrastructure with the purpose to provide to manufacturers zero defect solutions (Fig. 1).

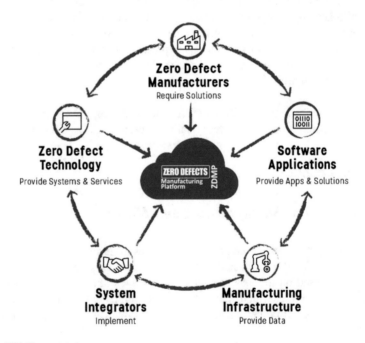

Fig. 1 ZDMP ecosystem

2 Process Quality Assurance Under the Zero Defect Manufacturing Scope

As stated in [2], ZDM implementation can be done according to a product-oriented perspective and a process-oriented perspective. The difference is that a product-oriented ZDM studies the defects on the actual parts and tries to find a solution, while, on the other hand, the process-oriented ZDM studies the defects of the manufacturing equipment and based on that can evaluate whether the manufactured products are good or not.

This paper focuses on ZDMP Process Quality proposal, focused on ensuring out-standing process quality through equipment, resource and energy efficiency by deploying novel AI-based solutions. Thus, based on the supporting services provided by the ZDMP platform, the process quality chapter will provide solutions addressing:

- **Start-up Optimisation**: Appling machine learning algorithms linked to part-flow simulation models and machine sensors to detect and correct configuration errors and anomalies to the setup and retooling of machines.
- **Material and Energy Efficiency**: Components detecting anomalies in the consumption which can also infer likely future-related defects. By identifying anomalies in use, preventative measures can be applied to the affected work centres and workpieces.
- **Equipment Optimisation**: Regression models to detect and take corrective measures to avoid machines making out-of-tolerance parts. By learning the relationship between process parameters, product properties, and quality, it allows actions on the equipment configuration to be promoted to avoid the occurrence of defects.
- **Process Quality Assurance**: Assuring the process quality and to make the manufacturing process self-adaptive. By building models of the process from other components with suitable configuration, it can adapt the optimisation goals and focus decisions on the best actions to optimise overall process quality and reduce unplanned downtime.

ZDMP will provide domain-specific services granting the development of zApps (ZDMP applications) with quality-specific models and algorithms that will be customised for the application-specific context.

3 Process Quality Assurance Microservices

ZDMP architecture is based on the principles of flexibility and composability and as such is based around an SOA and microservices approach [4]. As stated at [5], microservices are an architectural style for developing applications from the combination of microservices a business capability, which communicate with other microservices in an application through lightweight mechanisms. With this purpose, all ZDMP components implement and publish REST interfaces allowing

Fig. 2 Process quality components

the exchange of data (primarily) with a messaging bus. ZDMP supports event-driven SOA features so that the different components can decide their interaction pattern and react to internal and external events. Following this approach, the components of ZDMP can behave either as services and/or as event producers and consumers.

ZDMP is based on a federated architecture [6], based on IIRA model [7] and RAMI 4.0 [8].

The process quality assurance component in vf-OS is composed of three services (see Fig. 2), to be known: process prediction and optimisation designer, process prediction and optimisation runtime, process assurance runtime and process digital twin.

This paper covers the process prediction and optimisation designer, a component used by zApp developers for building process quality solutions based on machine learning pre-trained models and process optimisers. The three main principles of the solution are as follows:

– Machine learning and optimisers as a service [9], where AI development complexity is lowered, through the provision of AI-based solutions targeting specific needs in vertical industries and build sophisticated models to find actionable information with remarkable efficiency.
– Serverless architecture [9] as the architectural principle supporting AIaaS (AI as a service) and OaaS (Optimisers as a Service). Serverless computing introduces large-scale parallelism, and it was specifically designed for event-driven applications that require to carry out lightweight processing in response to an event.
– Machine learning (ML) algorithm selection [10], or optimiser algorithm selection [11] according to different criteria, for instance, in the case of ML, the accuracy versus interpretability. In the optimiser case, criteria are precision and the speed of computation.

The process prediction and optimisation designer will follow a set of steps driven by the machine learning project pipeline which answer will drive the selection of a subset of microservices available to solve a given problem and its configuration

Fig. 3 Prediction and optimisation designer workflow

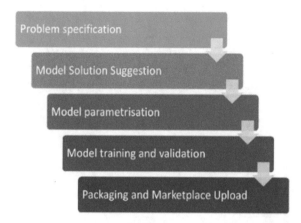

before deploying it to the process prediction and optimisation runtime. The stages are as follows:

- Identify the nature of the problem to be solved: manufacturing setup stage, process performance, resource consumption or other process quality assurance initiatives.
- Define a specific objective function defining the purpose at hand and expressed as the maximisation or minimisation of factors such as energy, scrap, waste of resources or lead time.
- Identify input and output data from a set of data models already pre-established.
- Specify non-functional priorities on factors such as time constraints, accuracy and interpretability.
- Select a specific algorithm, in case of being a supervised ML algorithm prepare training and testing data and train it. If non-supervised or optimisers are selected, only drive specific configuration of the algorithm.
- Evaluate performance of the algorithm and pack the model.
- Deploy the model as a serverless microservice on the process prediction and optimisation runtime (Fig. 3).

The packetised algorithm can be uploaded to the ZDMP marketplace, to be deployed on specific runtime instances of ZDMP and consumed by zApp microservices-based applications. In this final packetisation step, value-adding decisions are made. The first one, if the trained model is to be deployed along with the packetised algorithm. This model will save time when deploying the app in the production line, because minor training will be necessary, and some standardise problems can benefit from it. The second decision is the deployment model to consume the solution. In this sense, a process quality solution is a microservice offering endpoints to zApps that can be deployed to be run on a monolithic runtime or that will be run on distributed heterogeneous scenarios where services need to bring their own self-contained runtime stack. More details on this are provided in the next section of the paper.

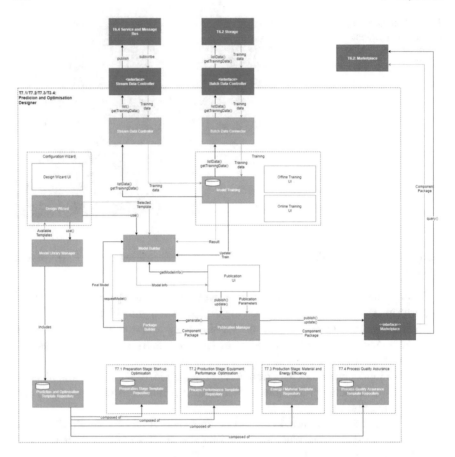

Fig. 4 Prediction and optimisation designer

The following schema (see Fig. 4) represents the internal structure and the connectivity of the process prediction and optimisation designer.

Among the main sub-components are

- **Prediction and Optimisation Template Repository**: This repository contains templates for optimisation, machine learning models and analytic techniques. Templates are linked to process optimisation or process quality prediction problems they are well suited for, subdivided in their corresponding optimisation domains:

  – Preparation Stage Template Repository: containing templates solving prediction and optimisation problems in the domain of process preparation stage (e.g. product changeover, process start-up).

- Process Performance Template Repository: This repository contains templates related to prediction and optimisation of manufacturing equipment performance.
- Energy/Material Template Repository: This repository contains templates to solve prediction and optimisation problems in the domain of energy and material efficiency.
- Process Quality Assurance Repository: This repository contains templates to solve process quality assurance problems.

- **Configuration Wizard**: Allows users to select the right template for a specific prediction or optimisation problem (e.g. minimisation of resource consumption, maximisation of production efficiency, prediction of CO_2 footprint) and to configure its parameters and data sources.
- **Model Training**: Generates scripts to update and train the model according to the information provided by the user and sends the script to the AI, Analytics Designer engine via the AI, Analytics Designer interface, so that the model is updated.
- **Stream Data interface**: Interface to the publish-subscribe functions of the message bus.
- **Batch Data Controller**: This module acts as a connector to list available data sources and receive batch training data sequences.
- **Model Builder**: Builds an executable script according to the information contained in the template and the input provided by the user.

4 Process Quality Microservices. Deployment Scenarios

As mentioned in the previous section, process quality models are microservices providing endpoints consumed by zApps. An example is a machine setup configurator model based on deep learning algorithms. These microservices are queried by zApps that can contain interfaces and transaction logic that consume the microservices to provide a production-ready solution.

Deployment of microservices solutions permits sharing data between distributed components managed by different companies or benefiting of cloud solutions lowering deployment complexity and easing their adoption by manufacturers.

4.1 Cloud-Edge Microservices Scenario

The hybrid cloud is the combination of public and private cloud that in this scenario are feature rich (meaning not computationally limited) on terms of hardware, networking and storage. This is common on companies with processes done on distributed locations with specific confidentiality needs that want to share and process

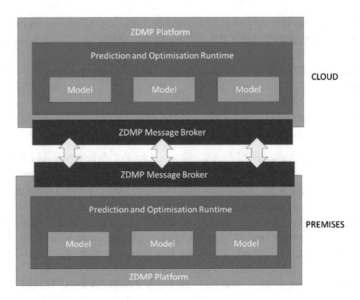

Fig. 5 Microservices on the hybrid cloud

information. In this case, two or more ZDMP Process Prediction and Optimisation Runtime Platforms run in parallel. Each of those runtimes will provide a stack of technology ranging from security, task management and training modules. Process quality model microservices will run and share the same stack of the runtime server that will be powerful but not suitable for low-power devices or appropriate for simple tasks. Figure 5 shows the schematics of this approach.

4.2 Distributed On-Premises Microservices

Self-contained microservices are runtime-stack-complete solutions intended for cases where limited resources are available, or easiness of a solution is necessary to limited complexity or scope of the problem to be solved at hand.

This architecture solution is common on supply chain scenarios where one company keeps a ZDMP platform and the main processes to be optimised, while other provider companies only want to share information with the first one, in the least intrusive way. An alternative business context is when low-powered devices want to preprocess data before sharing information to a functional-complete platform.

In this case, one ZDMP Process Prediction and Optimisation Runtime Platform run along with several stand-alone process quality model microservices. Each of those microservices makes minimum necessary processing before pushing information into the ZDMP platform. Figure 6 shows the schematics of this approach.

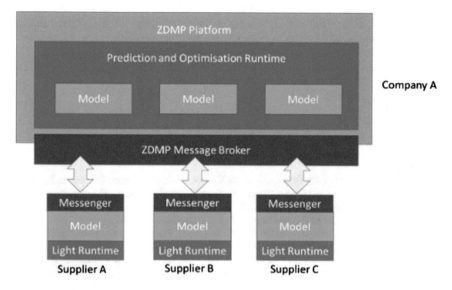

Fig. 6 Distributed on-premises microservices

5 Conclusions

This article has presented the ZDMP approach for providing AI as a service (AIaaS) and optimisers as a service (OaaS) for process quality assurance for zero defect manufacturing. The approach is fundamental as part of a broader ZDMP platform where basic core services (such as storage, data gathering or process engine) along with process and product quality assurance microservices are composed into zero defects application. The ZDMP process quality assurance is composed of four components. A prediction and optimiser designer and runtime in charge of loading, configuring, training, validation, deploying and running process quality models. The other two components are the digital twin virtualising and simulating processes and the quality assurance runtime reusing algorithms from the other two components for predicting and preventing defects.

The article has explained the prediction and optimisation designer as the core component to reusing and configuring machine learning existing models and optimisers and the key component that will pack and deploy trained models into serverless microservices. Two actual deployment scenarios have been presented. Those alternatives are not about technological concerns but about empowering the adoption of feasible solutions while respecting privacy and competences of the companies running cross-organisational quality preventive processes.

Acknowledgements ZDMP is a EU project funded by the H2020 Framework Programme of the European Commission under Grant Agreement 825631 and conducted from January 2019 until December 2022. ZDMP aims at providing an extendable platform for supporting factories with a high interoperability level, to cope with the concept of connected factories to reach the goal of zero defect production. ZDMP engages 30 partners (Users, Technology Providers, Consultants and Research Institutes) from 11 countries with a total budget of circa 16.2M€.

References

1. Wang, Z. Y., Qiu, Y. L., & Gui, S. H. (2006, March 1). Quality competence: A source of sustained competitive advantage. *The Journal of China Universities of Posts and Telecommunications, 13*(1), 104–108.
2. Psarommatis, F., May, G., Dreyfus, P. A., & Kiritsis, D. (2020, January 2). Zero defect manufacturing: State-of-the-art review, shortcomings and future directions in research. *International Journal of Production Research, 58*(1), 1–7.
3. ZDMP Homepage. https://www.zdmp.eu. Last accessed 2019/11/10.
4. Soldani, J., Tamburri, D., & Van Den Heuvel, W. (2018). The pains and gains of microservices: A systematic grey literature review. *Journal of Systems and Software, 146*, 215–232.
5. Li, J. (2020). Get Ready for the Emergence of AI-as-a-Service. THW blog article. Last accessed on January 2020 at https://thenextweb.com/podium/2020/01/24/get-ready-for-the-emergence-of-ai-as-a-service/
6. Fraile, F., Sanchis, R., Poler, R., & Ortiz, A. (2019). Reference models for digital manufacturing platforms. *Applied Sciences, 9*(20), 4433.
7. Industrial Internet Consortium. (2017). The Industrial Internet of Things Volume G1: Reference Architecture.
8. Deutsches Institut für Normung e. V. (2019). Reference Architecture Model Industrie 4.0 (RAMI 4.0) English Translation of DIN SPEC 91345:2016–04.
9. Pérez, A., Moltó, G., Caballer, M., & Calatrava, A. (2018, June). Serverless computing for container-based architectures. *Future Generation Computer Systems, 1*(83), 50–59.
10. Lee, I., & Shin, Y. J. (2019). Machine learning for enterprises: Applications, algorithm selection, and challenges. *Business Horizons*, November 26, 2019.
11. Andres, B., Poler, R., Saari, L., Arana, J., Benaches, J. V., & Salazar, J. (2018). Optimization models to support decision-making in collaborative networks: A review. In *Closing the Gap Between Practice and Research in Industrial Engineering 2018* (pp. 249–258). Springer, Cham.

A Declarative Approach for Change Impact Analysis of Business Processes

Adeel Ahmad, Henri Basson, Mourad Bouneffa, and M. Matsuda

Abstract The business process models provide a means to control and visualize the enterprise processes. Different processes in an enterprise inter-operate to achieve a common strategic and operational objective. These processes continuously evolve to meet the changing business requirements. In this respect, the process models should be able to reflect a cost-effective solution for the decided changes in a process and its impact on other executing processes. Such dynamic adaptability requires not only an exhaustive comprehension of business process activities but also the understanding of the various change dimensions. In this work, we propose a formal description of change feasibility, change incorporation, and traceability of the change impact propagation among multiple processes. A rule-based approach is proposed for change incorporation during the development and instantiation of business process models. The rule-based declarative approach is destined to estimate the change feasibility in dynamic business process models. We attempt to analyze the multiple dependency levels to better control the change impact propagation. The work aims to help a well-controlled and successful evolution of business processes.

Keywords Business process modeling · Business process evolution · Change impact analysis · Rule-based change management · Structural dependencies · Data dependencies

A. Ahmad (✉) · H. Basson · M. Bouneffa
LISIC, Université du Littoral Côte d'Opale, EILCO, Calais, France
e-mail: adeel.ahmad@univ-littoral.fr

H. Basson
e-mail: henri.basson@univ-littoral.fr

M. Bouneffa
e-mail: mourad.bouneffa@univ-littoral.fr

M. Matsuda
Kanagawa Institute of Technology, Atsugi, Japan
e-mail: matsuda@ic.kanagawa-it.ac.jp

© The Author(s), under exclusive license to Springer Nature Switzerland AG 2023   169
B. Archimède et al. (eds.), *Enterprise Interoperability IX*, Proceedings of the I-ESA
Conferences 10, https://doi.org/10.1007/978-3-030-90387-9_15

1 Introduction

Business process models (BPM) follow a continuous cycle of process discovery, process modeling, deployment, execution, improvement, and redesign [1, 2]. However, it is generally observed that the enterprises are reluctant to change the existing BPMs [3–5] because of the associated complexity and the cost. Indeed, the evolution of inter-operable business processes can generate difficult situations for the creation, modification, or deletion of process fragments in the rectified schemas. This problem can further aggravate when the instances of concerned process fragments are already in execution while introducing the change. It is because of the compliant of business process instances with the definition of their types, i.e., whether a respective change can correctly propagate its impact without causing inconsistencies or errors (e.g., deadlocks, live-locks) [6]. This can result a non-compliance with regulations [7] or a degradation of the quality of the business process [8, 9].

The changes at process instance level (also known as instance-specific changes) are often applied in an *ad hoc* manner to deal with the exceptions (unanticipated situations) resulting in an adapted instance-specific process schema [10]. These are specific to a particular instance, which means changes in one instance usually do not affect other running process instances. In many cases, changing the state of a process instance is not sufficient for a successful BPM evolution; the process structure itself has to be adapted as well [11]. For this reason, the change at the process-type level (also named as process schema evolution) is necessary to deal with the evolving nature of process roles (e.g., to adapt them to new legal requirements or new policies). The schema evolution often leads to the propagation of respective changes to the rest of the schema components and also to the ongoing process instances. This is particularly true if the instances have a longer runtime (e.g., medical or handling of leasing contracts).

The rest of the paper is structured as follows: Sect. 2 provides a brief overview of the related work. We explain, in detail, the dependency relationships and their analysis in Sect. 3, whereas Sect. 4 describes the assessment of the change feasibility and the analysis of the impact propagation of dynamic changes with the help of rules. We briefly discuss implementation prototype in Sect. 5. Later in Sect. 6, we conclude the content of this article.

2 Related Work

The research on change management of business processes has been continuing to attract increasing interest from the industry and the scientific community in the last couple of decades. The major focus remained on integrating changes into business processes without affecting running instances. While, it is observed that an a priori analysis of the change impact is given less consideration.

Several approaches and paradigms [12–16] have been proposed to cope with the changing processes and their flexibility. In [12], the authors suggest an algorithm to calculate the minimal region affected by the changes that is based on Petri Nets. It attempts to identify the change regions to check the compatibility of workflow changes. In [13], authors discuss a formal approach based on the notion of process constraints called constraint-based flexible business process management. It has been developed to demonstrate how the specification of selection and scheduling constraints can lead to increased flexibility in process execution, while maintaining a desired level of control. Similarly, the authors in [14] propose a combination of a set of change patterns and seven *change support features* dealing with the process change. In this regard, *YAWL* [15] is an initiative based on formal foundations that shows significant promise in the support of a number of distinct flexibility approaches. Also *Declare* [16], in this regard, offers to examine the change; its declarative basis provides a number of flexibility features. Interestingly, it supports transfer of existing process instances to the new process model.

In [17], the author suggests a flexible modeling and execution of workflow activities based on a business meta-model. This approach supports dynamic changes such as adding or deleting activities, but requires that the activity is not in the running state when incorporating the change.

Apart from the work listed above, in [18] the authors attempt to analyze the dependency relationships that exist within a workflow. However, their focus has been constrained on modeling the workflow rather than on the change impact analysis, and most of the dependency relationships are confined to the structural dependencies, i.e., intra-dependency of activity or routing.

In [19], the author presents a framework to analyze four types of dependencies concerning the activities, roles, data, and actors. The objective of this framework is limited to use this analysis to generate a set of "transition conditions" which are deployed in a distributed process control. The work of authors in [19] is closely relevant to our proposition. It uses the dependency analysis for the purpose of change impact analysis and suggests using a set of queries defined in PROLOG[1] to help designers and business experts to understand the dependencies between different elements of the business process model.

The use of rules makes the approach more general compared to the algorithms. We believe the declarative rules can help to determine the feasibility and assess an a priori change impact in multiple business process modeling languages (e.g., BPMN, EPC, UML activity diagrams).

3 Analysis of Dependency Relationships

We attempt to establish a scalable base to progressively consider the different interdependent dimensions of process models such as activities, data, actors, resources,

---

[1] http://www.gprolog.org/.

etc. Our objective is to identify the potentially affected elements for an a priori change impact analysis in the evolving business processes ahead the change implementation. We should consider the critical dependencies that may exist between the process model artifacts such as activities, data, roles, actors, resources, events, services, and rules, etc. In this paper, we specifically focus more on the multi-dimensional business process dependency model to get an insight concerning different dependency relationships among business processes.

In the following, we formally discuss some of the major dependency relationships in business processes.

3.1 Activity Dependency (Routing)

The activity dependency reflects the execution order of the business process activities. This ordering is usually defined by the modelers or business experts. It is based on technical requirements, legal regulations, and management policies. For example, if two activities are executed sequentially, it means that the completion of the execution of the first activity is a *pre-condition* for the execution of the second.

The activity dependency shows the execution order of activities within a business process through the control flows, i.e., *sequence flow* and *message flow*. This dependency defines not only the execution order but also the semantics associated with this ordering. For example, for an *AND-Join* routing of three-activities A, B and C; A and B must be executed before C (furthermore, in synchronization either A or B must finish before the C can start its execution, etc.).

An activity dependency is formally defined as: $D_a = (D_p, \Omega)$ over a set of activities $A = \{a_1, a_2, a_3, ..., a_n\}$ and a set of control flows $T = \{t_1, t_2, t_3, ..., t_n\}$, where:

$$D_p = D_p i(a) \ U \ D_p o(a), \ where \ a \in A \tag{1}$$

The $D_p i(a)$ is a set of all preceding activities $a_i \in A$ (denoted as: $a \rightarrow a_i$) on which the execution of activity a is dependent. The relationship can be a *many-to-one*, i.e., one activity depends on multiple activities.

In the same way, $D_p o(a)$ is a set of all succeeding activities $a_i \in A$ (denoted as: $a_i \rightarrow a$) meaning their executions depend on the activity a. The relationship can be *one-to-many*, i.e., multiple activities depend on one activity. The set of control flows (Ω), involved in the activity dependency, can be formally shown, as below:

$$\Omega = \Omega_i \cup \Omega_o \tag{2}$$

The $\Omega_i = (D_p i(a), a)$ is a set of control flows, $t_i \in T$, connecting each activity $a_i \in D_p i$ to a, i.e., all incoming arcs of a.

The $\Omega_o = (D_p o(a), a)$ is a set of control flows, $t_i \in T$, connecting a to each activity $a_i \in D_p o$, i.e., all outgoing arcs of a.

3.2 Role Dependency

The role is a logical abstraction of one or more actors, usually in terms of common responsibility or position. It means an actor can be a member of one or more roles. It is observed that a role is always associated with some activities.

The role dependency can be described through a role-net, which can be achieved by replacing the roles to the activities associated with them. In other words, the activity-based flowchart becomes a role-based flowchart while at the same time dependency relationships depend on routing entities.

For further clarification of the role dependency, let us consider a role R_1, which can be assigned to the same activities that are being executed by another role R_2, then the role R_2 may have a dependency relationship with the role R_1.

For the sake of further clarity, let us consider the activity *"blood test"* in a review process for medical checkup. It can be performed by a nurse or doctor. That is, the *role (nurse, blood test, medical checkup)* and *role (doctor, blood test, medical checkup)* are assigned to same activity which is *"blood test"*. Therefore, there exists a dependency relationship between the role *nurse* and the role *doctor*.

If we consider $R = \{r_1, r_2, r_3, \ldots, r_n\}$ as a set of roles and $A = \{a_1, a_2, a_3, \ldots, a_n\}$ as a set of activities, then the role dependency can be formally represented as:

$$D_r = (\sigma, \Psi) \text{ where, } \sigma(r) = \sigma_i(r) \cup \sigma_o(r), \text{ and } r \in R \qquad (3)$$

The $\sigma_i I$ represents the set of roles which are immediate predecessors of role r, i.e., these are the roles which are affected to the activities a_i where the activities $a_i \in A$ precede the activity a (for r associated to a). The $\sigma_o(r)$ represents the set of roles which are the immediate successors of role r. These are the roles which are affected by the activities a_i where the activities $a_i \in A$ succeed the activity a (for r associated with a). The set of control flows (Ψ), involved in the role dependency, can be formally shown, as below:

$$\Psi = \Psi_i \cup \Psi_0 \qquad (4)$$

The Ψ_i is the set of control flows ($t_i \in T$) or the arcs related to each role of $\sigma_i(r)$ to the role r, i.e., the set of incoming arcs of role r. In the same way, the Ψ_o represents the set of control flows ($t_i \in T$) linking the role r to all the roles of the $\sigma_o(r)$, i.e., the set of outgoing arcs of role r.

4 Declarative Assessment of Change Impact

In the presented approach, as broadly described in Fig. 1, the impact propagation is assessed with the help of rules written in *ECA* or <Event> <Condition> → <Action> formalism. It encompasses two steps, which are explained below:

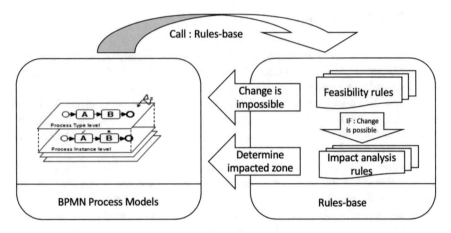

Fig. 1 Change impact assessment with the help of rules

1. Assess the feasibility of the dynamic change in BPMN process model with the help of a set of rules called feasibility rules.
2. If the change is feasible, then perform an a priori analysis of the impact propagation at the process-type level and in the corresponding instances with the help of a set of rules called impact analysis rules.

The change operations can be a combination of addition, deletion, or modification of activities, but these can also become more complex depending on their abstraction and granularity. The complex change operations can involve the replacement of a process fragment by another one, moving a process fragment from its current position in the flow to a new one, copying a process fragment, swapping a process fragment with another, parallelization of process fragments, or some other complex action. The meta-model of change impact analysis, as shown in Fig. 2, encompasses the possible prospects of the change. This provides a useful overview of the different concepts concerning the change and types of impacts to support the business process change impact analysis. Any change of a business process can propagate a multi-faceted impact, i.e., structural, functional, behavioral, logical, and qualitative impacts. Therefore, it leads to a comprehensive analysis as required by its definitions.

In the following, we formally describe the change impact analysis in business processes. A change operation can consequently result in a difference (denoted as Δ), between the initial process schema S_i and the modified process schema S_{i+1}. This can be expressed as follows:

$$S_{i+1} = S_i + \Delta \tag{5}$$

$$\Delta = |S_{i+1} - S_i| \tag{6}$$

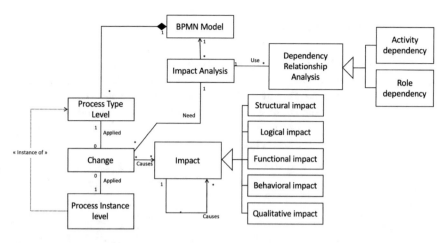

Fig. 2 Meta-model of change impact analysis

The variant (Δ) can generate the post-change impacts on whole or part of the process model and its running instances. Therefore, an a priori analysis of this variant is important to ensure the correctness and consistency of the change impact propagation. Otherwise, changes such as the deletion or the addition of a task may cause severe inconsistencies (e.g., unintended update loss) or even run-time errors (e.g., program crashes due to the invocation of task modules with invalid or missing parameters).

4.1 Feasibility Rules

The set of *Feasibility Rules (FR)* ensures the compliance of business process instances to the definition of their type during a change. It can be used to assess the feasibility of the dynamic changes. To further illustrate, let us consider, as described below, the example of a process-type level change.

The rule *process-type level change* ensures the feasibility of the dynamic change at process-type level. It is defined as follows:

- In order to avoid the insertion of a new task T as a predecessor of an already *RUNNING* or *COMPLETED* task, we require that all the succeeding elements in the control flow must be in one of the states as *NOT_ACTIVATED* or *ACTIVATED*. Conversely, the preceding tasks may be in an arbitrary state.
- The deletion of a task T of a running process instance is only possible, if T is either in *NOT_ACTIVATED* or in *ACTIVATED* state. In this case, the elements associated with T are removed from the corresponding process model. Tasks in the *RUNNING, COMPLETED*, or *SKIPPED* state may not be deleted (it should not be allowed to delete a task or to change its attributes if it is already completed).

4.2   Process Instance Level Changes

A feasibility rule at the process instance level can be triggered to control the changes
at the process instance level. We instantiate process graphs, where the set of nodes
can be either activities, events or gateways. The set of sequence flows (edges) connect
the nodes. Let us consider *status* as an attribute assigned to each node N and each
instance I to describe its current status (change-trace). The Algorithm 1 describes
such a rule for the sake of illustration.

Algorithm 1 Deletion of process fragment.

```
on Iₓ is < deleted >
if Iₓ ∈ S then

    I ← Inst( Iₓ );
    /* Verification of corresponding instances */
    if I ∈ {Not_Activated, Activated} then
    /* the change can be applied */

        Status( Iₓ, "deleted");
        Mark( Iₓ, GREEN);

    else

        /* the change cannot be applied*/
        print("the change cannot be immediately
        applied to the "+ Iₓ +" instance");

    end if

end if
```

4.3   Impact Analysis Rules

When the change is possible, impact analysis rules (analyze the impact propagation)
are triggered, such as described in the Algorithm 2.

As shown in the Algorithm 2, the FO_x and other relevant control flows are marked.
The set D_po returns both succeeding activities and the corresponding routing relationships of the given activity and returns succeeding activities (if multiple activities
depend on concerned activity) in respect to the different routing types: Sequential,
AND-Split, AND-Join, OR-Split, OR-Join, XOR-Split, XOR-Join.

The set D_pi returns both preceding activities and the corresponding routings
between preceding activities and the given activity, respectively (if the concerned
activity depends on multiple activities). All returned activities and corresponding

Fig. 3 Deletion of activity "B" and addition of activity "X"

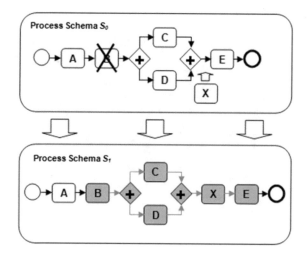

routing relationships (control flows) are also marked to express the depth of change impact, such as shown in Fig. 3, with the help of an example.

Algorithm 2 Change impact analysis (activity dependency)

```
on Rule_06 is < called >
if Status(FOx)== "added" ‖ "deleted" ‖ "modified"
then

    Mark(FOx, BLUE);
    Mark(FC ∈ {Ωi(FOx) U Ωo(FOx)}, BLUE);
    /* Dpo gets successively each succeeding
    activity depending on FOx and the corresponding
    routing relationships in N */
    for ai ∈ N(i = 1,...,n) do

        if ai → FOx then /* ai depend on FOx */ Dpo ← DpoU{ai}
        end if

    end for
    for ai ∈ Dpo(i = 1,...,n) do

        Mark(ai);
        Mark(FC ∈ {Ωi(ai) U Ωo(ai)});

    end for
    /* Dpi gets successively each preceding
    activity, on which FOx depends and
    the corresponding routing relationships in N */
```

```
for aᵢ ∈ N(i = 1, . . . ,n) do

    if FOₓ → aᵢ then /* FOₓ depend on aᵢ */ Dₚi ← Dₚi ∪ {aᵢ}
    end if

end for
for aᵢ ∈ Dₚi (i = 1, . . . ,n) do

    Mark(aᵢ);
    Mark(FC ∈ Ωᵢ(aᵢ) U Ωₒ(aᵢ));

end for

end if
```

5 Prototype of Validation

The proposed approach has been validated with the help of plug-ins development in Eclipse[2] integrated development environment. Among others, we have been developing a BPMN Change Propagation Analyzer plug-in to extend the functionality of BPM modeling for traceability of impact propagation for changing business processes. The plug-in is composed of the management of meta-information of business processes and a rule base allowing the implementation of the rules developed in context of analyzing the change integration and impact propagation.

The rule base is implemented using the Drools[3] object-oriented rule engine in integration with Java[4]. This rule engine allows the management of business process change impact propagation rules. Indeed, Drools is a business rules management platform that offers an integrated rule definition and execution workshop. It also allows the definition and execution of the workflow as well as the management of events. The set of impact propagation rules is interactively called when handling BPMN template elements (add, delete, or modify).

6 Conclusion and Perspectives

In this paper, we propose to analyze the change impact by exploiting the dependency relationships between BPM elements. In this respect, we focus on activity, data, and role dependencies among business processes. The approach

---

[2] The Eclipse Foundation—IDE and tools. https://www.eclipse.org/.

[3] Drools—Business Rules Management System. https://www.drools.org/.

[4] Java. https://www.java.com/.

is based on graph reachability with the help of a rule-based framework. The feasibility rules and the change impact analysis rules are the two major categories of rules in this regard. These can effectively determine an a priori feasibility and analysis of process changes either at process-type level or process instance level.

The approach has been validated with the help of set of plug-ins which are developed for Eclipse IDE. The continuing work aims to analyze the change impact propagation on the multiple dependency relationships, which include actors, resources, events, control data, and applications in the business process on both at the process-type level and the process instance level. The rule-based approach may provide an assistance to assess the feasibility and change impact analysis for better business process management.

References

1. Bouneffa, M., & Ahmad, A. (2013). Change management of BPM-based software applications. In *Proceedings of the 15th International Conference on Enterprise Information Systems*, Angers, France, July 2013. SCITEPRESS.
2. Dumas, M., Rosa, M. L., Mendling, J., & Reijers, H. A. (2013). *Fundamentals of business process management*. Springer Publishing Company.
3. Recker, J., & Mendling, J. (2016). The state of the art of business process management research as published in the bpm conference. *Business and Information Systems Engineering, 58*(1), 55–72.
4. Reijers, H. (2006). A comprehensive approach to flexibility in workflow management systems. In *Proceedings of the 15th IEEE International Workshops on Enabling Technologies: Infrastructure for Collaborative Enterprises (WETICE 2006)*, 26–28 June 2006, Manchester, United Kingdom (pp. 271–271). IEEE Computer Society.
5. Reijers, H., & van der Aalst, W. (2005). The effectiveness of workflow management systems: Predictions and lessons learned. *International Journal of Information Management, 25*(5), 458–472.
6. Kherbouche, O., Ahmad, A., & Basson, H. (2012). Detecting structural errors in BPMN process models. In *Proceedings of the 15th IEEE International Multitopic Conference (INMIC)*, Islamabad, Punjab, Pakistan (pp. 425–431). IEEE Computer Society.
7. Awad, A., Decker, G., & Weske, M. (2008). Efficient compliance checking using BPMN-Q and temporal logic. In *Proceedings of the 6th International Conference on Business Process Management* (pp. 326–341). Springer.
8. Snchez-Gonzlez, L., Ruiz, F., Garca, F., & Piattini, M. (2013). Improving quality of business process models. In *Proceedings of the 6th International Conference, ENASE 2011*, Beijing, China (Vol. 275, pp. 130–144). Springer.
9. Ahmad, A., Basson, H., & Bouneffa, M. (2019). For a better assessment of business process qual ity. *DEStech Transactions on Computer Science and Engineering*, ITEEE.
10. Minor, M., Schmalen, D., Koldehoff, A., & Bergmann, R. (2007). Structural adaptation of work-flows supported by a suspension mechanism and by case-based reasoning. In *Proceedings of the 16th IEEE International Workshops on Enabling Technologies: Infrastructure for Collaborative Enterprises Cover* (pp. 370–374). IEEE Computer Society.
11. Reichert, M., Dadam, P., & Bauer, T. (2003). Dealing with forward and backward jumps in work-flow management systems. *Software and System Modeling, 1*(2), 37–58.
12. Sun, P., & Jiang, C. (2009). Analysis of workflow dynamic changes based on petri net. *Information and Software Technology, 51*(2), 284–292.

13. Rinderle, S., Reichert, M., & Dadam, P. (2004). Correctness criteria for dynamic changes in work-flow systems a survey. *Data and Knowledge Engineering, 50*(1), 9–34.
14. Weber, B., Reichert, M., & Rinderle, S. (2008). Change patterns and change support features enhancing flexibility in process-aware information systems. *Data and Knowledge Engineering, 66*(3), 438–466.
15. van der Aalst, W., & ter Hofstede, A. H. M. (2005). Yawl: Yet another workflow language. *Information Systems, 30*(4), 245–275.
16. Rosa, M. L., Van Der Aalst, W. M., Dumas, M., & Milani, F. P. (2017). Business process variability modeling: A survey. *ACM Computing Surveys (CSUR), 50*(1), 2.
17. Mendling, J., Weber, I., Aalst, W. V. D., Brocke, J. V., Cabanillas, C., Daniel, F., Debois, S., Ciccio, C. D., Dumas, M., Dustdar, S., et al. (2018). Blockchains for business process management-challenges and opportunities. *ACM Transactions on Management Information Systems (TMIS), 9*(1), 4.
18. Huang, Y., Huang, J., Wu, B., & Chen, J. (2017). Modeling and analysis of data dependencies in business process for data-intensive services. *China Communications, 14*(10), 151–163.
19. Dai, W., & Covvey, H. D. (2005). Query-based approach to workflow process dependency analysis. School of Computer Science and the Waterloo Institute for Health Informatics Research, Waterloo, Ontario, Canada, Tech. Rep.

Model-Driven Approaches

A Reference Model for Interoperable Living Labs Towards Establishing Productive Networks

Majid Zamiri, João Sarraipa, and Ricardo Jardim Goncalves

Abstract Interoperability has been regarded as an influential factor, but (concurrently) critical for organizations and enterprises that intend to become powerful, efficient, and competitive in turbulent markets. Furthermore, open-innovation ecosystems and research environments such as Living Labs (LLs) have shown high potential capabilities in establishing, joining, and/or incorporating in an interoperable network of labs. Interoperability enables the LLs to exchange their knowledge, information, data, experiences, and findings with other labs at different levels. An interoperable LL on the one hand has the potential to raise its values, e.g., to promote the workflows and improve outcomes, and on the other hand to reduce the duplication of its efforts or errors. On this account, the main goal of this study is to propose a (generic) reference model for interoperable living labs (RM-ILL) that can steer and support LLs in establishing and developing interoperable network(s) of labs. RM-ILL can also facilitate the understanding of related concepts used in this specific context. RM-ILL is developed in the light of ARCON modelling framework which helps to achieve interoperability at different levels of abstraction and collaboration. This study demonstrates the applicability and usefulness of RM-ILL, addressing how RM-ILL can help the establishment and development of an interoperable network of labs supported by the CARELINK project.

Keywords Interoperability · Reference model · Living Labs (LLs) · Network

1 Introduction

The gradual evolution in collaborative practices and environments has opened a large set of opportunities for enterprises to for example benefit from interoperability standards, improve their information services, and gain knowledge from multiple sources. In that sense, interoperability (inside and outside the organizations) is becoming an indisputable reality in today's networked world establishment requirements. As such,

M. Zamiri (✉) · J. Sarraipa · R. J. Goncalves
Faculty of Sciences and Technology and UNINOVA—CTS, NOVA University of Lisbon,
2829-516 Caparica, Portugal
e-mail: ma.zamiri@campus.fct.unl.pt

© The Author(s), under exclusive license to Springer Nature Switzerland AG 2023
B. Archimède et al. (eds.), *Enterprise Interoperability IX*, Proceedings of the I-ESA
Conferences 10, https://doi.org/10.1007/978-3-030-90387-9_16

interoperability has already caused substantial growth in connectivity and transactions in almost every domain, where multiple parties with different perspectives can work together, drive better and timelier access to reliable information, and enhance their operational efficiencies which in turn can lead to better outcomes.

Interoperability is considered a key success factor for almost all branches of industry to survive and compete. Additionally, it can offer a number of benefits including but not limited to diminishing organizations' costs (e.g., of integration, operation, maintenance) and also increasing organizations' agility, competitiveness, efficiency, and stability [1, 2]. Even though interoperability is highly beneficial, it has some disadvantages and costs. It may, for instance, compromise the privacy and security, or add some technical complexity to the system design and also inflicts new requirements on a system.

Recent advances in understanding the basic mechanisms of interoperability led to the emergence of new trends and successful applications in different realms including, computer science, information system, healthcare, communication, etc. Interoperability can also be employed in open-innovation ecosystems such as LLs or living laboratories that are often operating in a territorial context (e.g., region, city) in order to integrate concurrent research and innovation within or between public–private entities. LL often relies on the combination of research and innovation and focuses on co-creation approaches. On that account, LLs evolve through exploration, investigation, experimentation, evaluation, and implementation of related concepts, scenarios, ideas, technologies, artifacts, and services in a real-life context. It should be noted that the stakeholders of LLs collaborate with each other through different methods and for different purposes (e.g. achieving an efficient system, product, or service to consequently promote the quality of life) [3]. This type of collaboration can be taken place at two levels, within a LL or between the LLs, actually and/or virtually. In the case of collaboration between LLs, when the number of LLs is increasingly growing and they are geographically distributed, they all can reap the advantages of networked interoperable LLs [4, 5].

Given the above, it is evident that the networking of existing and emerging LLs is now a key activity [6]. There is not, however, in the literature a comprehensive model that can conceptually and theoretically define and clarify the process of establishing and developing interoperability in LLs. Thus, this study aims to *propose the RM-ILL to guide establishing interoperable network(s) of LLs. The RM-ILL provides an abstract representation of the concepts and the main elements needed for networking of LLs.* The RM-ILL can assist developers and researchers in better planning, implementing, developing, operating, and maintaining the network of interoperable LLs. The rest of the paper is structured as follows: Section 2 describes the base concepts used in this study. Section 3 briefly describes the nature of interoperable LLs. Section 4 clarifies the proposed reference model for LLs and presents a demonstration scenario. The paper ends with short conclusions and future work.

2 Base Concepts

Interoperability—is a broad concept that is conceived on different levels of abstraction. Among the several proposed definitions of interoperability, some appear more commonly used than others. Generally, interoperability refers to the ability to of share knowledge, information, data, and experiences between some entities. Interoperability is in particular defined as an ability of two or more heterogeneous and autonomous operating entities such as organizations, enterprises, businesses, systems (IT applications, solutions and components), or people who collaboratively work with each other in physical and/or virtual environment to maximize the opportunities for sharing or reusing the knowledge, whether internally or externally. Interoperability describes the extent to which those entities can orchestrate a reliable delivery of knowledge, information, data, digital objects and/or resources via software and hardware in meaningful ways. Furthermore, interoperability not only allows organizations to harness the benefits of knowledge sharing and collaboration but also assists them in functionally link their activities in an efficient way. Therefore, organizations could and should be interoperable as it facilitates collaboration, knowledge management, and also increases their openness, adaptability, and productivity [6]. As illustrated in Fig 1, interoperability between two organizations/entities such as LLs can be taken place at three levels: (a) knowledge (semantic) level, (b) Business (organizational) level, and (c) ICT (technological) level [7]. In this study, the focus of attention is on the business (organizational) level.

Living Lab—the term "Living Lab" is at risk of becoming a buzzword in the research, innovation, and collaboration domain, since it lacks a consistent or commonly accepted definition. Moreover, a vast range of activities, indeed, can be performed through LLs, and each, in turn, has its own requirements and conditions.

According to [9, 10], LL can refer to one or more of the following attributes:

- It is an open-innovation ecosystem and research centre for sensing, prototyping, validating, and refining a variety of solutions by involving multiple stakeholders.
- It is a space that by taking the advantages of knowledge sharing can be used for developing a project, product, service, and/or system.
- It is a participatory methodology relying on collaboration between various agents of a system.
- It involves end-users in the product development process.

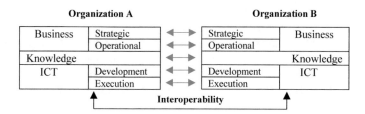

Fig. 1 Levels of interoperability between two organizations/entities [7, 8]

Fig. 2 Key components of a
Living Lab

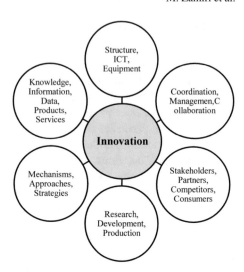

Despite a commonly agreed definition is now beyond our reach, a better understanding can be gained by finding what constitutes the basis of a LL environment. Hence, from our view, a LL needs the following key components exhibited in Fig. 2.

As a type of open and flexible laboratory, the LLs are basically set up to generate ideas, develop research and experiments around the ideas, then validate the findings and make them documentation, and lastly provide (new) products or services. This process occur when the right stakeholders in a LL (e.g. researchers, designers, innovators, entrepreneurs, developers, academics, associations, SMEs, and also consumers) with the needed competencies come together at the right time to create added value and deliver something new collectively. Each stakeholder can take the advantage of provided opportunities in different ways. For instance, organizations not only can get the new ideas but they can increase the return on their investments, researchers can extend their research and development activities, and consumers can get better and new products and services. In addition, a multitude of businesses has reported that interoperability not only helps them to provide better products and services, but it makes shorter the time of market development [7].

3 Interoperable Living Labs

Over the years, the nature and morphology of LLs have been constantly changed and redefined. The force that drives the trends towards increasing openness and interactions has turned the attentions to interoperability and collaborative efforts, both internally and externally which empower LLs to not only promote their efficiency, but also benefit from various resources. Furthermore, interoperability assists LLs in being more visible, usable, compatible, and prosperous. Evidence shows that organizations (in general) and LLs (in particular) are not originally interoperable. In order an organization or a LL transforms to an interoperable organization or lab, in addition to e.g., the needed entities, rules, tools, and methods, it also requires harmonized workflows to avoid unexpected and accidental disruptions. To that end, a suitable model can provide a clear roadmap and needed directions for the specific executions. The model, besides, can help to present the related issues in a more structured way. The literature [11, 12] addresses a variety of requirements that should be taken into consideration for establishing and developing interoperability. For organizations, it is essential to achieve interoperability at all three before-said levels. With that into account, the most commonly addressed requirements for three levels of interoperability are presented in Table 1.

Taking Table 1 into account as general requirements for interoperability and also identifying the needed specifications for each particular LL, an interoperable LL can then move towards establishing and developing a network of interoperable LLs by considering the following proposed activities [13]:

Phase I (building the LL):

- Identifying the main stakeholders and bringing them together,
- Setting the scope, vision, and goals of the LL,
- Providing required environment and equipments,
- Identifying a related reference model, e.g., LLRM (see Sect. 4),

Phase II (designing, adapting, and developing the reference model):

Table 1 Some most common requirements for interoperability in three levels

Interoperability requirements		
Knowledge/semantic ensures organizations understand the meaning of exchanged knowledge as intended	Business/organizational ensures processes, responsibilities and expectations for exchanging knowledge are aligned across all relevant stakeholders	ICT/technological ensure platforms, systems, and applications can exchange or process knowledge
• Developing information management strategy • Providing tools to control the use of terms and language • Using metadata standards and schemas • Using data quality reporting out- puts • Developing and maintaining enterprise data models	• Aliening business process • Aligning requirements to similar organizations • Developing organizational relationships • Understanding how data assets are used to meet organizations' outcomes • Designing and delivering services for stakeholders • Implementing positions	• Using whole-of-Government platforms • Selecting machine readable file formats • Using application programming interfaces • Providing knowledge exchange services • Using technologies to transform and improve legacy knowledge • Using standardized knowledge exchange specifications

- Designing and adapting an appropriate reference model for the LL,
- Validation and endorsement of the reference model,
- Developing the reference model,
- Implementing the reference model, and
- Measuring the outcomes.

For implementing and improving the interoperability, the LLs need to assess qualitatively and quantitatively their capabilities, compatibility, performance, and etc. [14]. Whenever the needed requirements for interoperability are fulfilled and the LL turned into an interoperable LL, the LL can then make links with other similar and interoperable LLs around it. Through creating active and effective network(s), the LLs can collaborate to (better) achieve their personal, common, or compatible goals as well as access wider/new resources. Such network(s) can provide an opportunity for almost all stakeholders to harness the benefit of a collaborative learning environment. For example, the European Network of Living Labs (ENoLL) in November 2006 setup a sustainable network to develop and offer a gradually growing set of networked LL services. This network aims to make the innovation in the industry more efficient and dynamic by involving the stakeholders in the development of new products, services, and societal infrastructures [15].

3.1 Living Labs in the CARELINK Project

The CARELINK is an Active and Assisted Living (AAL) project and is co-financed by the European Commission (through Horizon 2020) and by internal countries' budgets until 2020. The CARELINK aims to create a better quality of life for elderly people who suffer from dementia. AAL is an area where artificial intelligence (AI) can play an essential role, in particular, in the support of ageing people. The CARELINK project came up with the objective to help people with dementia (PwD) (who may affect by physical, emotional, and economic constrains), their caregivers, and families [10].

CARELINK by networking the related LLs in the healthcare domain for the PwDs and also by taking the advantages of experience and knowledge sharing attempt to create an impact on the development of created solutions. That is, in those LLs, through exchanging the participants' experiences, expertise, knowledge, feedback, workloads, findings, and also by benefiting from combined views of different minds, the CARELINK tried to make positive changes on the process of product design, development, and specifications as well as on the commercial approach taken. For this purpose, the CARELINK project created a partnership with a successful LL named "Internet of Things Open Laboratory (ITOL)" [10]. In this journey, the RM-ILL provided useful support and directions. This issue is elaborated in Sect. 4.1.

4 Reference Model for Interoperable Living Labs (RM-ILL)

There are a variety of suggested definitions for the reference model. From collaborative network point of view, a reference model is an abstract representation of a model pattern or domain-specific ontology that synthesizes and organizes the core concepts, constituting elements, and practices for a particular network/entity. In order for a LL to join interoperable networks and capture the complexity of this process, the RM-ILL is proposed. Among other available modelling frameworks in the literature (e.g., Zachman Framework, SCOR, VERAM, EGA, FEA), the RM-ILL is inspired from A Reference Model for Collaborative Networks (ARCON). The ARCON is a framework that helps to develop reference models for any types of collaborative networks. ARCON clarifies how the heterogeneous elements of various collaborative network are gained from three distinct viewpoints namely, the environment characteristics perspective, the model intent perspective, and the life cycle perspective toward making a potential reference model.

Relying on our understanding and findings, we believe that the ARCON reference model in comparison with the above-mentioned modeling frameworks is more suitable for the nature of collaborative networks (such as networked interoperable LLs) in terms of scope, framework, and target. However, it is worth mentioning that "the reference model is generic and not directly applicable to concrete cases. It rather provides the basis for an organized derivation of other specific models closer to these concrete cases" [16].

It should be also noted that in the application or adaption of ARCON (particularly for LLs), a number of stakeholders sould be considered as main participants including researchers, experts, educators, decision-makers, producers, innovators, and developers. Despite the successful applications that the ARCON has already had, it is still evolving and reaching its near-maturity stage. Thus, it needs further investigation, application, development, and validation. It is note taking that ARCON is already used in several contexts (e.g., in collaboration networks for information improvement of food consumers [17]) as well as it is applied in different industries, services, and research projects. For example, ARCON "is initiated in the framework of the European ECOLEAD project, it will then be developed by international organizations such as SOCOLONET (Society of Collaborative Networks) and IFIP (International Federation for Information Processing) WG 5.5 (COVE: Cooperation Infrastructure for Virtual Enterprises and electronic Business") [18].

Taking into account the above-mentioned prerequisites (addressed in Table 1) for interoperability, *the RM-ILL (by inspiration from ARCON reference model and adapting it for interoperable LLs) tries to address two major environment characteristics, namely: (a) internal elements characteristics (labeled endogenous elements) and (b) external interactions (labeled exogenous interactions) that address the surrounding environment and interactions with outside the network of LLs.*

The endogenous elements embrace four main dimensions, including:

- *Structural dimension* addresses the structure and components of the network (e.g., participants, their relationships, and network typology). Furthermore, this dimension handels the roles that will be taken by the participants in the network.
- *Componential dimension* focuses on tangible and intangible resources of the network (e.g., different resources such as software and hardware resources, human elements, knowledge, and information). Additionally, this dimension also deals with network´s ontology and the flow of knowledge and information at different levels.
- *Functional dimension* addresses the base functions and operations of the network (e.g., processes and procedures that are associated with different phases of the networked collaboration life cycle).
- *Behavioural dimension* presents the principles, policies, and governance rules that drive the behaviour of the network.

The exogenous elements similarly consist of four main dimensions, including:

- *Market dimension* covers the issues related to the interactions with customers and competitors.
- *Support dimension* focuses on those support services that are provided by the third-party entities (outside the network).
- *Societal dimension* addresses the issues related to the interactions between the network and society in general.
- *Constituency dimension* presents the interactions with the potential new members of the network (e.g., interactions with those entities that are not yet part of the network but might be interested to contribute).

The proposed endogenous elements for LLs are addressed in Table 2, and the exogenous elements are presented in Table 3. These two Tables are the main output of this work. They present a general reference model that has potential application to different types of LLs. However, for each specific type of LL, the RM-ILL should be initially adapted according to, for example, the purpose of application, requirements of the environment, and qualification of participants.

As addressed in Table 3, three main groups of elements are considered for exogenous elements, including:

- *Network identity* defines the environment in which a LL is positioned, (it shows the position of LL in the environment, and addresses the way in which a LL presents itself in the environment)
- *Interaction parties* identify the potential entities that LL can interact with
- *Interactions* list the type of transactions that a LL can develop with its interlocutors.

Table 2 Endogenous elements for LLs

Endogenous elements for networked interoperable LLs			
Structural dimension	Componential dimension	Functional dimension	Behavioural dimension
Participants – They are volunteer – They have diverse profession and background – They are distributed Roles – Coordinator – Administrator – Support provider – Researcher Roles relationship – Supervision – Collaboration – Communication – Knowledge sharing – Socializing – Peer to peer – Trusting Network typology – Network has access criteria – Network size is unlimited – Knowledge flows in two sides (transmit and receive)	Resources • **Domain-specific devices** – Equipment – Tools • **Technological resources** – Hardware – Software – Internet • **Human resources** – Multi-stakeholder Users Public Academia Firms • **Knowledge resources** – Participants' profile data – Networks' profile data • **Network outcomes** – Findings – Tangible products – Services	Processes • **Fundamental processes** – Network management Membership management Profile management Role management Execution management Trust management • **Background processes** – Network management Repository creation Management system setup Ontology management Execution evaluation Procedures • **Fundamental processes** – Objective setting – Rules setting – Requirement provision – Network creation – Technology adoption – Registration – Role assignment – Participation – Interaction – Quality assurance – Conflict resolution – Risk management – Knowledge management	Governance model – User-driven Rules and Policies – Joining needs permission – Initial training is needed – Active participation is a key – Value creation – Conflict resolution – Content validity is a must – Trust is advised – Support is encouraged – Network requires protection – Collaboration development Agreements – Long/short term contribution – Terms of participation – Findings ownership – Acknowledge findings

Table 3 Exogenous elements for LLs

Exogenous elements for networked interoperable LLs

Network identity

Market dimension	Support dimension	Societal dimension	Constituency dimension
Mission – Limitless network research – Collaborative learning Network profile – Virtual LL networks – Connection building by online platforms Market strategy – Learning service – Research service – Partnership creation	Network's social nature – LL is open-innovation and user-centred system – LL is collaborative multi-contextual environment – LL is impinged by the direct influence and actions of surrounding stakeholders – LL can join/cooperate with private/public sectors – LL can be launched for profit/non-profit purposes – LL can be extended to real-life environment	Status – Networked interoperable LL is an informal entity Values – LL cultivates public learning – LL fosters collective research and development – LL promotes collaborative and innovative production	Attracting and recruiting strategies: – Visibility increasing – Flexibility increasing – Easy approaches to inclusion and exclusion – Updating related information – Advertising the networks, vacancies, and findings – Conduct workshops

(continued)

Table 3 (continued)

Exogenous elements for networked interoperable LLs

Market dimension	Support dimension	Societal dimension	Constituency dimension
Interaction parties			
Customers	• **Financial entities**	• **Public organizations**	• **Public organizations**
• **Strategic customers**	– Banks	– Healthcare institutes	– Educational centres
– Public/Private organizations	– Investors	– Insurance companies	– Libraries
– Businesses	– Sponsors	– Social security	– Laboratories
– Producers	• **Technical entities**	– Social services	– Clinics
– Educational centres	– IT companies	– Civil defence	– Welfare services
– Research centres	– Technical experts	– Media institutions	– Public Charities
– Strat ups	– Engineers	• **Private sectors**	• **Business organizations**
• **Potential customers**	– Network service provider	– Educational organizations	– Manufacturers
– Related laboratories	– Storage service provider	– Training services	– Instructors
– Problem solving markets	• **Informational entities**	– R&D organizations	– Merchandises
– Knowledge services	– Universities	– Manufacturers	– Service businesses
– Individuals	– Libraries	– Producers	– SMEs
Competitors	– Professional enterprises	– Brokers	– Traders
– Similar LLs	– Research institutes	• **NGOs**	• **Private organizations**
– R&D networks	– Researchers	– Education charities	– Teaching services
– Innovative research labs	– Experts	– Market advocacy NGOs	– Learning services
– Knowledge services	• **Social entities**	– Technical assistance NGOs	– Emergency services
– Research services	– Public/Private organizations	– Environmental NGOs	– Instructors
Suppliers	– Charities	• **Interested entities**	– Career centres
– Equipment producers	– Individuals	– Producers	• **Research organizations**
– Equipment service/repair	• **Training entities**	– Farmers	– Research councils
– Libraries	– Professional associations	– Consulting services	– Science services
– Education centres	– Advisors	– Training institutes	– Science parks
– Training services	– External experts	– Supporters	• **Technology centres**
– Consult services	– Other recognized labs	– Workshops	– IT services
– Experts		– Educators	– ICT professionals
		– Authors	– Innovative centres
		– Vendors	– Technical schools
			– Innovative centres
			– Engineering centres
			• **Sectors/Individuals**
			– Civic sector
			– End-users

(continued)

Table 3 (continued)

Exogenous elements for networked interoperable LLs

Market dimension	Support dimension	Societal dimension	Constituency dimension
Interactions			
Customers transactions – Co-creation by consumers and end-users – Enquiries handling – Consulting Competitors transactions – Partnering – Integrating – Dealing Suppliers transactions – Collaboration – Supporting	• **Service acquisition** – Financial support – Technological support – Information service – Consulting service – Training service – Guarantee service – Donation service • **Agreement establishment** – Integrating – Partnership – Dealing – Sponsoring – Network affiliation	• **Political relations** – New/Wider collaboration with potential organizations and people • **Social relations** – Cultural relations – Contract relations – Patronage • **Learning** – Public awareness – Public learning – Public research – Public production • **Seeking support:** – Knowledge and experience exchanging	• **Member searching** – Advertising/calling – Joining is encouraged and supported – Invitation sending – Participants can bring in new faces – Current participants should be maintained • **Joining mechanism** – **Applicant:** sends application form – **LL:** checks the application, and: Accepts the application, Rejects the application or Requests correction

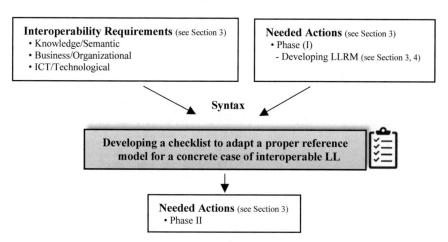

Fig. 3 Main inputs for developing a checklist to adapt an appropriate reference model for a concrete case of interoperable LL

4.1 Demonstration Scenario

The RM-ILL provides a picture of the core components of a typical interoperable LL. The RM-ILL can be used as a guide for creating and adapting an appropriate reference model for a concrete case of interoperable LL (e.g., a LL for people dealing with dementia and wandering). In this direction, both the above-mentioned interoperability requirements (shown in Table 1) and also the needed actions for building a LL (pointed out in Section 3, phase I) can be used as a base, directly or indirectly. Fig 3 illustrates the main inputs (interoperability requirements and need actions) for building a checklist that is useful for adapting the reference model for the target interoperable LL.

The CARELINK project has started the creation of an interoperable LL in order to support the dissemination and exploitation of the project results that can impact the society particularly the PwD. The CARELINK project in cooperation with ITOL applied the RM-ILL to enable a further smooth establishment of LL productive networks. In addition, by taking the advantages of RM-ILL, the authors try to use the gained knowledge and experiences by this work and apply them in the SHYFTE project [10] and build a network focusing on Skills 4.0 LABs.

In the following, an excerpt of a sample checklist (based on Fig 2) is presented in Table 4. This checklist is a prerequisite for interoperable LLs consideration prior to join a productive network of LLs. The checklist helps LLs to be better prepared for the needed qualifications and specifications. It can also be served as a guide for adapting an appropriate reference model for a concrete case of interoperable LL. As an example, checklist number 2.2 relates to the goals of a LL that intends to become interoperable and join to a network of interoperable LLs. In the case of the CARELINK and ITOL, the goal was to help the installation configuration and

Table 4 Sample of checklist to adapt a reference model for a concrete case of interoperable LL

Sample checklist

No.	Inputs	Considerations	Yes	No
1	**Interoperability requirements**			
1.1	Knowledge	Are information management strategies developed?		
		Are the tools to control the use of terms provided?		
		Are the metadata standards and schemas used?		
1.2	Business	Is business process aliened?		
		Are requirements to similar organizations aligned?		
		Are organizational relationships developed?		
1.3	ICT	Are whole-of-Government platforms used?		
		Are machine readable file formats selected?		
		Are application programming interfaces used?		
2	**Needed Actions (Phase I)**			
2.1	Main stakeholders	Are they identified and brought together?		
		Are they mutually agreed to collaborate?		
		Are they maintain a shared understanding of the tasks?		
2.2	Goals	Are all goals clearly set?		
		Are the goals properly specified?		
		Are the goals periodically adjusted?		
2.3	Environment	Is the needed environment provided?		
		Has the environment standard conditions?		
		Is it adapted for physical and virtual collaboration?		
2.4	Reference model	Is the related reference model identified?		
		Are all its endogenous elements considered?		

(continued)

Table 4 (continued)

Sample checklist

No.	Inputs	Considerations	Yes	No
		Are all its exogenous elements considered?		
3	**Needed Actions (Phase II)**			
3.1	Adopting a proper reference model	Is accordingly a proper reference model adapted?		
		Are all its endogenous elements well adjust?		
		Are all its exogenous elements well adjust?		
3.2	Validation	Are the input parameter values considered?		
		Is the provided information examined?		
		Is the expert judgement applied?		
3.3	Designing	Are the concepts refined and mapped to the application?		
		Are the milestones created?		
		Are the indicators defined?		
3.4	Implementation	Is implementation plan created?		
		Are the required standards defined?		
		Are the risks and issues identified and recorded?		
3.5	Measurement	Are the responsible entities for measurement specified?		
		Are the measures determined?		
		Are the measures determined?		

the use of devices that work with low power wide area networks (LPWANs) such as cellular-based technologies like LTE-NB1 (also referred to as NB-IoT) [10].

5 Conclusion and Future Work

Developing feasible solutions for different levels/classes of interoperability (technical, organizational, and semantic) is going on. In this regard, the proposed reference models in the literature each attempt to offer an appropriate suggestion for different aspects of organization interoperability (e.g., related concepts, components, and interaction). This research work proposes the RM-ILL (based on ARCON modelling framework) for LLs in order to enhance the understandability of the related concepts, to use the model as a consolidated basis for further developments, and for the purposes of discussion among researchers, educators, developers, and producers. The RM-ILL tries to conceptualize, in the highest level of abstraction, the environment and the main characteristics of interoperable LLs. It also attempts to envisage the external interactions between a LL and its surrounding area which can be used as a guide for establishing productive network(s) of interoperable LLs. Therefore, the RM-ILL aims to streamline the design and development of a particular model for a concrete case of LL by providing generic solutions. In this respect, the authors of this study have started the building and developing a specific LL and a network of LLs (for people dealing with dementia). In future work, the authors will first develop the checklist that could be used for maturing the RM-ILL. Next, the focus will be given to the application of RM-ILL to other potential projects and use cases.

Acknowledgements The authors acknowledge the project CARELINK, AAL-CALL-2016-049 funded by AAL JP, and co-funded by the European Commission and National Funding Authorities of Ireland, Belgium, Portugal, and Switzerland.

References

1. Soon-Yong, Ch., & Whinston, A. B. (2000). Benefits and requirements for interoperability in the electronic marketplace. *Technology in Society, 22*(1), 33–44.
2. GridWise, Architecture Council. (2009). Retrieved from: https://www.grid-wiseac.org/pdfs/fin ancial_interoperability.pdf
3. Rodrigues, M., & Franco, M. (2018). Importance of living labs in urban entrepreneurship: A Portuguese case study. *Cleaner Production, 180*, 780–789.
4. Jara, C. A., Candelas, F. A., Torres, F., Dormido, S., Esquembre, F., & Reinoso, O. (2009). Real-time collaboration of virtual laboratories through the Internet. *Computers & Education, 52*(1), 126–140.
5. Desai, K., Jin, R., Prabhakaran, B., Diehl, P., Belmonte, U. H. H., Ramirez, V. A., Johnson, V., & Gans, M. (2017). Experiences with Multi-Modal Collaborative Virtual Laboratory (MMCVL). In *Third international conference on multimedia big data*. IEEE, Laguna Hills, CA, USA.
6. Sten-Erik, Ö. (2017). *Interoperability capability to interoperate in a shared work practice using information infrastructures: studies in ePrescribing.* Doctoral thesis, Department of Management and Engineering, Linköping University, Sweden.
7. Ralyté, J., Backlund, P., Kühn, H., & Jeusfeld, M. A. (2006). Method chunks for interoperability. In: *25th international conference on conceptual modeling*, Tucson, AZ, USA.
8. Sarraipa, J. (2013). *Semantic adaptability for the systems interoperability.* PhD Thesis, Faculty of Science and Technology, New University of Lisbon.
9. Higgins, A., & Klein, S. (2011). Introduction to the living lab approach. In Y. H. Tan, N. Björn-Andersen, S. Klein, & B. Rukanova (Eds.), *Accelerating global supply chains with IT-innovation.* Springer.
10. Zamiri, M., Marcelino-Jesus, E., Calado, J., Sarraipa, J., & Goncalves, R. J. (2019). Knowledge management in research collaboration networks. In *International conference on industrial engineering and systems management (IESM)*, Shanghai, China, Sep 25–27. https://doi.org/10.1109/IESM45758.2019.8948162

11. Mallek, S., Daclin, N., & Chapurlat, V. (2012). The application of interoperability requirement specification and verification to collaborative processes in industry. *Computers in Industry, 63*(7), 643–658.
12. The Law Library. (2018). *Interoperability requirements, standards, or performance specifications for automated toll collection systems (US Federal Highway Administration Regulation) (FHWA).* CreateSpace Independent Publishing Platform.
13. NHS England, HSCIC, South, Central and West Commissioning Support Unit. (2019). Interopera bility Handbook.
14. Leal, G., Guédria, W., & Panetto, H. (2019). Interoperability assessment: A systematic literature review. *Computers in Industry, 106*, 111–132.
15. Mulder, I., Fahy, C., Hribernik, K. A., Velthausz, D., Feurstein, K., Garcia, M., Schaffers, H., Mirijamdotter, A., & Ståhlbröst, A. (2007). *Towards harmonized methods and tools for living labs.* eChallenges.
16. Camarinha-Matos, L. M., & Afsarmanesh, H. (2007). A comprehensive modeling framework for collaborative networked organizations. *Intelligent Manufacturing, 18*(5), 527–615.
17. Volpentesta, A. P., Felicetti, A. M., & Frega, N. (2019). Collaboration networks for information empowerment of food consumers. In L. Camarinha-Matos, H. Afsarmanesh, & D. Antonelli (Eds.), *Collaborative networks and digital transformation. PRO-VE 2019. IFIP advances in information and communication technology* (vol. 568). Springer.
18. Ermilova, E., & Afsarmanesh, H. (2008). Further steps on CN reference modeling. In L. M. Camarinha-Matos, & H. Afsarmanesh (Eds.), *Collaborative networks: reference modeling.* Springer.

Integrated Model-Based Configuration of Production Systems—Reflection of ISO 19440 and MDA and MDI

Thomas Knothe, Jan Torka, Patrick Gering, and Frank-Walter Jäkel

Abstract Rising business competition leads to complexity because of increased number of product variants and customer-specific processes. Model-based approaches seem to be suitable for handling this kind of flexibility in networked production environments. In this paper, current approaches to the configuration of heterogeneous systems based on standard models are reflected, and an integrated model-based configuration approach using formalized modules is proposed and its application demonstrated.

Keywords Enterprise modelling · Modular architectures · Model transformation heterogeneous production systems

1 Introduction

1.1 Business Challenge—Order-Specific Process Implementation

In the ongoing competition in the industry, the individualization of products is becoming more and more important [1, 2] and [3]. In order to meet the require- ments of its customers, production is constantly shifting from series production to individual and small series production. Thus, the producing companies are more and more confronted with the situation of producing in batch sizes of up to 1 and more, which means creating order-specific procedures that even produce the same product [4]. This is the case if the law in a customer's country restricts certain parts coming from a banned country.

The demand for individualized products is prompting companies to rethink their existing production concepts and try to make them more flexible. In particular, they must be prepared for changes in their production resources by adding, replacing or

T. Knothe (✉) · J. Torka · P. Gering · F.-W. Jäkel
Fraunhofer IPK Berlin, Pascalstraße 8-9, 10587 Berlin, Germany
e-mail: Thomas.Knothe@ipk.fraunhofer.de
URL: http://www.ipk.fraunhofer.de

© The Author(s), under exclusive license to Springer Nature Switzerland AG 2023
B. Archimède et al. (eds.), *Enterprise Interoperability IX*, Proceedings of the I-ESA
Conferences 10, https://doi.org/10.1007/978-3-030-90387-9_17

deleting items beyond the usual configuration processes [5]. In order to meet the requirements of the market, production systems must be able to be adapted quickly to the new requirements [6]. Today's production systems, which are designed and optimized for rigid line production, do not meet the requirements and can only be adapted to the new situation with a high amount of time and resource consumption. Jovane predicted as early as 2003 that customer order-specific processes would also be required in the former series business [7].

In the next chapter, a given use case is provided and based on an evaluation model for the derivation of a sustainable digitization strategy [8] reflected according to the required capabilities of the entire production system environment.

1.2 Application Case

The given example of an order-specific production deals with the automated machining and handling gears and assembling into a gearbox (Fig. 1).

The scenario includes the following assets:

- Product: gearbox and gears individualized by different kind of machining operations,
- Technical hardware resources:

  - Two robots for handling, assembly operations and polishing, using different effectors like grippers and polishing spindle,
  - Controlling units for robotic operations
  - Force sensors at polishing and grippers, temperature sensor, optical sensor,
  - Shopfloor IT-system hardware,

- Technical software resources to be involved:

  - Modular software system applying a model-based execution engine,

Fig. 1 Cycloidal gearbox and part of production environment

- Web-based product configurator,
- Shopfloor management system for order handling,
- Cockpit,

- Processes: Value stream for handling and polishing gears and finally the gear-box assembly and control process to trigger production steps and handling feedback from resources.

The case requires the third level of digitization: Digitalized individual operations according to the mentioned evaluation model [8]. Therefore, all corporate assets have to be enabled at this level too.

Human resources and partner issues are not covered in the application case and will not be taken into account furthermore. The approach for this case is to have a complete configurable set of production environment which can handle order-specific processes, based on customer demands. The individual customer demand has to lead to a specific product configuration (gear size and surface specification). The product data have to be used to generate individualized manufacturing processes, which are managed by an order handling system. Further on the control processes and parameters have to be adjusted and observed in real time according to the specific demands.

To enable the corporate assets mentioned in this paper, a concept is presented utilizing an enterprise model-based approach for a modular execution architecture for handling order-specific processes. In the next chapter, business, methodological and technical requirements are defined to cover the mentioned business case. In Chap. 3, current standards for enterprise modelling and architectures for model-driven development are adapted to these requirements. In the following chapter, a model-based approach is presented, in which the experiences of applications are described.

2 Requirements

The environment as defined above has to enable the order-specific process paradigm by enabling:

- Process logic changes on demand according to the customer's specification, not only for the product but also for the business and technical processes. The adaptations have to be made within a typical customer cycle time.
- Changes have to be understandable by machining experts, plant managers and IT specialists.
- The hardware and software systems described above have to be interoperable through flexible interaction and adaptive configuration regarding the required changes.

In the following figure, the case is assigned according to the smart manufacturing evaluation model [1]. This open model identifies four levels of flexibility, ranging from digital transparency of corporate structures to integrated operations. The cases were assigned to the individualized operations according to the business requirements. This means the entire company network of assets must be built up at least statically, while product, process and resource structures have to be established dynamically.

Taking into account the business requirements mentioned above, the following essential methodical and technical requirements are derived from an enterprise model-based approach:

1. **Expressiveness of modelling constructs:** The modelling constructs used to perform the changes must be sufficiently expressive to allow multilateral collaboration between the different roles mentioned above.
2. **Covering customer-specific requirements:** The modelling constructs must cover all customer-specific requirements as well as the process and system adaptations in order to perform the changes.
3. **Immediate change:** Changes to the defined specifications have to be transferred immediately and without data loss into an executable format and environment.
4. **Execution environment:** A solution must include an execution environment, capable of executing the changes.
5. **Parallel change and execution:** The configuration of assets has to be made possible in parallel of ongoing business operations.

Most of these requirements are mentioned a long time ago, e.g. in [9]. Nevertheless, the trend towards increasingly flexible production is becoming more and more relevant for the actual business. In the next chapters, current standards in enterprise modelling and model-based interoperability will be reflected to cover the mentioned business, methodological and technical requirements.

3 Reflection of Current Standards

Among standardization, there are several standards dealing with enterprise modelling and their applications. These are:

- ISO 15704:2000: Industrial automation systems—Requirements for enterprise-reference architectures and methodologies [10]
- ISO 19439:2006: Enterprise integration: Framework for enterprise modelling [11]
- ISO/FDIS 19440: Enterprise modelling and architecture—Constructs for enterprise modelling [12]

ISO 15704 describes a framework to integrate system life cycle, view types and enterprise assets according to the GERAM approach (generalized enterprise reference and methodology). ISO 19439 is focusing on the CIM-OSA framework and the consequences for the modelling approach according to the four major views:

function, information, resource and organization. ISO 19440 defines the generic architecture of modelling constructs, their standard properties and their connection in a model.

Model-driven architecture (MDA) describes a model-driven framework for software development developed by the object management group (OMG) [13]. The architecture split the complexity for developing and adapting executable systems into three connected model types: Computer-independent model (CIM), platform-independent model (PIM) and platform specific model (PSM) which is close to being executed. The model-driven interoperability (MDI) can be seen as a derivation of MDA to enable seamless interoperability between heterogeneous systems. It is maintained by the organization I-VLAB.

Regarding flexible execution environments, several approaches have been developed in the last ten years, e.g. "Industrie 4.0 component" for flexible involvement of products, services and processes. The work to complete is still ongoing, e.g. in the German Platform Industries 4.0, so their achievements are not reflected in this paper.

3.1 Standards Regarding Modelling

The GERAM approach of ISO 15704 provides a very good framework of terminology in enterprise modelling. The given system life cycle approach is comprehensive but applicable in a very general manner. For using this approach in an order-specific process paradigm, formal model-based interfaces are missing to enable a nearly automatic transfer between different phases of the life cycle. Without these relationships, a systems engineering approach would take too long time to implement the required adaptations.

The same applies to ISO 19439, where modelling concepts are very limited, e.g. the separation of functional, informational and resource views makes the integrated engineering of solutions components complicated (e.g. for adapter building blocks, which have all three views to be integrated). For both 15704 and 19439, it is not organized, how a parallel approach for customization and operation would work.

The reflection on ISO 19440 is concentrating to the requirements no 1 and 2 regarding Sect. 2 of this paper. The given modelling constructs are suitable to cover all aspects of the case, even some overlapping issues complicates the assignment (e.g. order and event constructs). The generic approach of properties makes it possible to address all specific data required for the case as described in Sect. 1.2. Because of the complicated meta-model in ISO 19440, it seems very difficult to facilitate the common understanding between different roles. Furthermore, ISO 19440 does not provide any representation rules and related guidelines for given constructs.

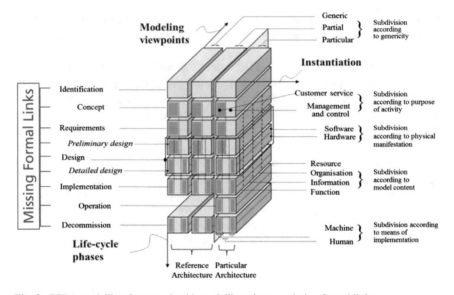

Fig. 2 GERA modelling framework with modelling views—missing formal links

3.2 Model-Driven Architecture (MDA) and Model-Driven Interoperability (MDI)

The development process of the MDA is divided into the same phases as the traditional approach. The decisive difference lies in the form of the recording between the individual phases. Instead of records in the form of diagrams and text descriptions, machine-readable models are created. The three core models of the MDA are described below [14].

The platform-independent model (PIM) is created during the analysis phase. The PIM describes a software that takes a very specific use case into account. It does not consider how this software is technically implemented, but only how the use case is best supported. The requirements as a starting point for this phase are still mainly defined in text form.

In the next phase, the platform-independent model is transformed into one or more platform-specific models (PSM). The PSM considers the specific technology of the implementation and needs knowledge of the specific platform to understand the model. Since today's systems usually consist of several technologies, there are usually several PSMs. The final phase of development is the transformation of the PSM into executable code. While the transition from one phase to the next in traditional software development is usually done manually, the advantage of MDA lies in the automatic transformation.

The model-based approach and the automatic transfer from PIM to PSM and PSM to code make it possible to exploit several advantages over the traditional approach (Kleppe et al. 2007):

Regarding the requirements mentioned in Sect. 2, MDA covers:

- Automatic approaches in the transformation from PIM to PSM and coding to enable immediate application of customer demand changes
- Interoperability between specific execution systems, because a PIM can be transferred to several PSM
- Existing tools supporting the maintenance of the models—parallel approaches of execution and change seem possible
- Execution of models by using UML-based execution systems
    There are still a couple of drawbacks in MDA:
- Insufficient automatic transformation between CIM to PIM—change on demand seems impossible
- UML-based approaches are still oriented on the level of computer scientist, for other roles they are too complicated
- A lot of different diagrams makes application on shop-floor level difficult
- Some of the given diagrams are not formal (e.g. use case diagram) and do not cover all aspects (e.g. interfaces)

The model-driven interoperability [15] is a specialization of MDA with the focus to enable interoperability between two enterprises from an engineering perspective. The major approach is to establish an interoperability model on each architecture level.

The advantage of this approach is the more flexible integration of systems coming from heterogeneous sources (e.g. a quality management system to perform parameter tracking along the production coming from the customer). The mentioned uncertainty issue with MDA is not solved, so that with MDI, the problems regarding change and execution on demand will be even greater because of the more interfaces and different models.

The following table (Table 1) presents the conclusion of the evaluation of related ISO standards and MDA/MDI.

The main shortcomings of the given approaches are:

- Manual transformation between different levels of abstraction and formalisms
- Limited applicability for the different stakeholders in networked companies
- Too many interfaces to manage between different models

To conclude, even if some aspects are sufficiently covered, the given approaches for model-based support of customer-specific processes are limited.

Table 1 Evaluation of standards and specifications

Requirements	ISO 15704, ISO 19439, ISO 19440	MDA, MDI
Expressiveness of modelling constructs	○	○
Covering customer-specific requirements	◓	◓
Immediate change	○	◓
Execution environment	○	●
Parallel change and execution	n.A	●

Legend: ○—not covered, ◔—partly covered, ◓—mostly covered, ●—completely covered

4 Integrated Model-Based Configuration of Execution Systems and Operations Control

By observing the major drawbacks of current approaches, a model-based solution methodology has to solve the dilemma of being suitable applicable for all stakeholder and at the same time formal enough for being executable in order to perform changes immediately. Therefore, three major decisions have been taken:

– Only one model transformation between design and execution model
– Design model has to cover both: being formal and useful for different stakeholders
– Applying an execution engine, covering all technical and business aspects

In Fig. 3 the basic building blocks for an architecture are indicated. The design model uses the "integrated enterprise modelling" which is at the same time in general understandable by all involved roles and compliant to ISO 19440. Here all aspects for executing the "move"-services are integrated. The modular architecture has to make sure that all required module combinations are possible and at the same time to avoid combinations which are forbidden. Each module is described according to a formal service description language. Here, the Unified Service Definition Language USDL is used. There is a formal mapping between the enterprise modelling language and USDL. Based on this constraint, the resulting enterprise model can be seen as formal to be executed in the execution engine. This execution engine is organizing the federation between the different connected execution systems.

The core USDL specifications are listed below [16]:

– **Foundational Module:** Set of concepts and properties, such as time, location, organization, etc. that are used in all modules.
– **Service Module:** General information about the service type, nature, titles, taxonomy and descriptions.

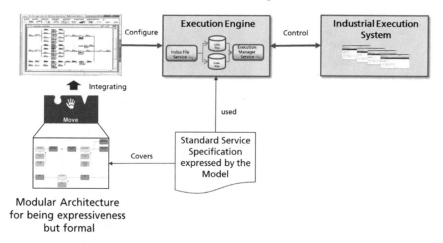

Fig. 3 Basic concepts for the integrated architecture to consider customer order-specific processes

- **Participant Module:** Participating organizations, contact persons and their role within the service fulfilment.
- **Functional Module:** Information about the specific capabilities of a service, input/output parameters and constraints.
- **Interaction Module:** Points of interaction and the responsible participants or participant roles in course of the service fulfilment.
- **Technical Module:** Mapped of functions (capabilities) of a service to technical realizations of the service (e.g. WSDL operations, parameters, faults, etc.)
- **Pricing Module:** Price plans, price components, fences, etc. for a service.
- **Service Level Module:** Service level agreements, such as time schedules, locations and other constraints.

In the following Fig. 4 and example of a value, process module is given based on IEM methodology [17]. Except the foundation module properties, all other modules are related to the basic constructs of IEM for the formal mapping to USDL. So USDL service is modelled as a resource being executed for the transport. For instance, this resource type is directly containing properties of service, pricing, service level and technical description.

The entire modular architecture provides for a limited number of modules formal models. The modules are separated into three different function types:

- Standardized interaction function
- Standardized basic information function
- Generic operation module

Generic operation modules combine execution functions on the field level. The standardized interaction functions are IT modules which are used for the in- and

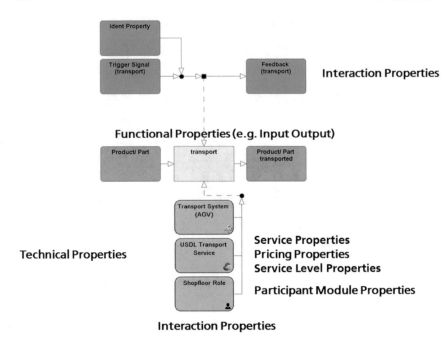

Fig. 4 Example module (transportation) based on IEM and the assignment of relevant USDL parts

output parameters from the module. Standardized basic information functions are used for other data processing, e.g. data synchronization (Fig. 5).

The functions "signal setting" and "handle feedback" are standardized interaction functions. A standardized basic information function is the function "data processing". "Mating" is an example for a generic operation module. With the help of the modules of the modular shop floor IT, different production processes can be mapped by combination or rearrangement. This enables a flexible reconfiguration

Fig. 5 Function types of modular architecture

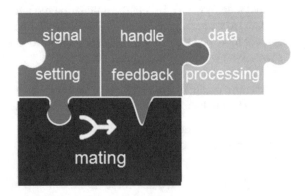

for customized products and variants as well as a faster setup of new production processes [18].

By using this model-based approach, complete execution environments can be configured (Fig. 6). For the given use case, the model has to be linked to the product configuration, to configure itself the order management, the production sequence monitor, production dashboard and the execution engine. The execution engine is triggering the automation assets like the robots and their effectors or the transportation systems. The execution engine is using an index file and the USDL provided by the model. So the model-based execution engine controls the production process. To do this, it executes the respective services in the order defined in the index file. For better administration, the services are combined in a central service registry.

For the integration, different concepts are applied. In the given case, the OPC UA framework is used as core interface. The adapter contains three different functions:

- Transformation of specific protocols to OPC UA
- Validation of OPC UA implementations on the equipment side by comparing expected configuration files according to given specifications against the data generated by the system
- Emulating real systems for integration tests

The direct logic connection between the enterprise model and the production equipment as well other execution systems (e.g. the dash board) requires on their side a change of services and its entire architecture.

Whilst in traditional robotic applications, kinematic procedures (movements) are dominating such architectures in the given case task oriented (e.g. drill a hole) modules have to be established. Here a synchronisation between process model and automation tasks has to be established.

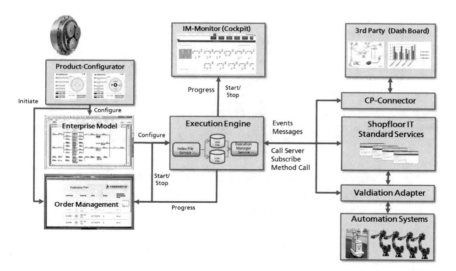

Fig. 6 Specific system architecture based on the basic concepts (see Fig. 3)

5 Application

In the given use case, 26 different variants of specific processes can be generated based on customer product and process requirements. The first step is the customer-specific specification of the product, which is carried out with the help of the product configurator. The customer selects the middle gear position for all three gear positions as an example and also wants to have them polished. The data is transferred to the database, and the customer order is created with the status "order created" and the specific customer data.

The second step is the automatic creation of the process model. In this case, the process model consists of six modules, three modules each for the insertion of the gearwheel and three modules for the polishing. With the help of a reference model, which contains a simplified description of all variants, and the module library, which contains all available modules, the control process model is created automatically (Fig. 7). The model is stored in the database including the existing-specific customer data. The model is then further transformed the index file and the various USDL files. These are also stored in the database, and the status of the customer order is changed to "process model created".

The third step is the execution of the production process by the model-based execution engine. As soon as the model-based execution engine no longer executes an order, and at least one order with the status "process model created" exists, all required order data is loaded from the database and the status of the order in the database is set to "process is being executed".

As soon as one service has been successfully processed, the next one is started. Parallel to the execution of the order, the current process list and the status of the individual processes are displayed in the dashboard. In addition, the time already

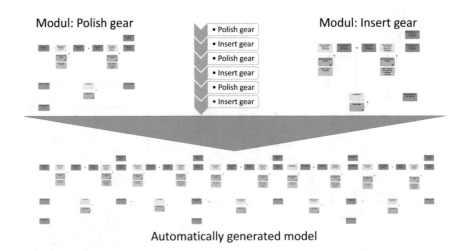

Fig. 7 Automatic integration of different modules into one specific production model

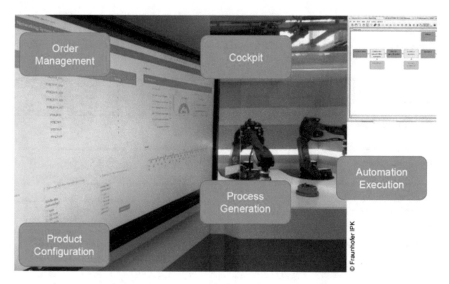

Fig. 8 Use case environment with connected product configurator, IM monitor, dashboard and order management system

spent is also displayed. As soon as the entire process is finished, the status of the order in the database is set to "order completed". The entire case were tested and executed by different roles and competences. In Fig. 8, the entire use case environment is shown.

All in all the following major experiences by applying the proposed modular architecture can be stated:

- The interoperability between different systems and across a system life cycle can be achieved by just one model transformation under the following constraints:

  - Formalized modular architecture
  - Mapping of a suitable service definition and execution language to the specialized modelling constructs (USDL to IEM)
  - Application of validation adapters
  - Harmonized set of business and technical function and process elements

- Order-specific processes can be generated in a suitable time frame
- The used IEM was applicable for all different roles
- Changes of processes for a new order can be generated in parallel to the current executed procedures
- Specific customer requirements which are not covered in the technical and process modules needs to be implemented manually. This is needed, e.g. in case of the integration of specific manufacturing tools and not foreseen handling tasks.

In the following table, the comparison between the standard approaches and the integrated model-based configuration of modular components is provided (Table 2).

Table 2 Evaluation of standards, specifications and the given approach

Requirements	ISO 15704, ISO 19439, ISO 19440	MDA, MDI	Integrated model-based configuration
Expressiveness of modelling constructs	○	◐	●
Covering customer-specific requirements	◐	◐	◐
Immediate change	○	◐	●
Execution environment	○	●	●
Parallel change and execution	n.A	●	●

Legend: ○—not covered, ◐—partly covered, ◑—mostly covered, ●—completely covered

Except the covering of customer-specific issues, all other requirements are fulfilled by the integrated "one-transformation" approach.

6 Conclusion

In this paper, an integrated model-based approach for interoperability of multiple connected different production systems and across system life cycle is proposed and its application demonstrated. The major aspects of this are: "one-transformation"-paradigm, modular architecture, complete formalized models but expressiveness for different stakeholders. This approach is reflected to the current standards in modelling and model-based information systems engineering and interoperability (MDA and MDI).

The experiences from application of the approach in the test environment at Fraunhofer IPK are demonstrating the intended benefits. Nevertheless, there are several draw-backs which have to be considered for industrial application. Most important is the needed harmonized business and technical function architecture and its implementation. In fact, this needs a lot of resources for reconfiguring and reorganization of PLC programming and its management. Therefore, a transformation approach is required.

From standardization point of view, the ISO family of ISO 19440, 19439 and 15704 needs to be revised according to the approach to deal with model transformation. Therefore, it would be suitable to extent ISO 19440 by application scenarios to demonstrate the usefulness of the constructs for standard cases. In total, MDA and MDI are still suitable for developing systems from scratch, but both should be extended to introduced how they can be used in case of modular formalized model artefacts.

References

1. Lindemann, U., & Baumberger, G. C. (2006). *Individualisierte Produkte—Komplexität beherrschen in Entwicklung und Produktion.* Springer.
2. Jovane, F., Westkämper, E. & Williams, D. J. (2009). The manuFuture road. Towards competitive and sustainable high-adding-value manufacturing. Springer.
3. Reinhart, G. & Zäh, M. (2003). Marktchance Individualisierung. 1 Hrsg. Springer.
4. Mörtl, M. (2008). *Ressourcenplanung in der variantenreichen Fertigung.* Utz.
5. Abele, E. & Reinhart, G. (2011). Zukunft der Produktion. Herausforderungen, Forschungsfelder, Chancen. Carl Hanser.
6. Kirchner, C., Seyfarth, M. & Wurst, K.-H. (2004). Modellbasiertes Rekonfigurieren von Werkzeugmaschinen, s.l.: wt Werkstattstechnik online.
7. Jovane, F., Koren, Y. & Boer, C. (2003). Present and future of flexible automation: Towards new paradigms. In CIRP annals—Manufacturing technology. s.l.:s.n., (pp. 543–560).
8. Knothe, T., Oertwig, N., Obenaus, M. & Butschek, P. (2018). Evaluation model for the derivation of a sustainable digitisation strategy. s.l., International Conference on Intelligent Systems.
9. Mertins, K., Knothe, T., & Zelm, M. (2004). User oriented Enterprise Modelling For Interoperability with UEML in EMMSAD'04, Riga Latvia, June 7–8, 2004.
10. International Standardisation Organization: ISO 15704:2000: Industrial automation systems — Requirements for enterprise-reference architectures and methodologies
11. International Standardisation Organization: ISO 19439:2006: Enterprise integration: Framework for enterprise modelling.
12. International Standardisation Organization: ISO/FDIS 19440: Enterprise modelling and architecture—Constructs for enterprise modelling.
13. Object Management Group: MDA®—The architecture of choice for a changing world. http://www.omg.org.mda. [Last access 5 11 2019].
14. Singh, Y., & Sood, M. (2010). The impact of the computational independent model for enterprise information system development. *International Journal of Computer Applications. 11* https://doi.org/10.5120/1602-2153
15. Ducq, Y. (2013). An architecture for service modelling in servitization context: MDSEA In M. Zelm, M. van Sinderen, L.F. Pires, & G. Doumeingts (Eds.), *Enterprise Interoperability: Research and Applications in Service-oriented Ecosystem (Proceedings of the 5th International IFIP Working Conference IWIE 2013)* (pp. P3–21). Wiley.
16. W3C Incubator Group. (2011). Unified Service Description Language XG Final Report. Available at: https://www.w3.org/2005/Incubator/usdl/XGR-usdl-20111027/ [Last access 5 11 2019].
17. Spur, G., Mertins, K., & Jochem, R. (1993). *Integrierte Unternehemensmodellierung.* Beuth.
18. Jaekel, F.-W., et al. (2018). Model based, modular configuration of cyber physical systems for the information management on shop-floor. In C. Debruyne, et al. (Eds.), *On the move to meaningful internet systems* (pp. 16–25). Springer.

A Usage Model to Enrich MDSEA Approach

Christophe Merlo, Véronique Pilnière, and Katarzyna Borgiel

Abstract An information system supports the actors' activities of an organization. Thus, the implementation of a new digital tool is a process of mutual transformation between organization and technology, involving changes on several dimensions. More specifically, its implementation has an impact on the practices of the actors, resulting in significant changes that can be "positive" and/or "negative" for the actors and their work activities. Our work intends to support these changes to improve interoperability projects based on MDSEA approach and encourage so-called positive changes, while facilitating the implementation of new IT solutions. We propose to understand both the variability of organizational change and the tool enhancements. Thus, we propose a usage model to characterize individuals' "usages" and their variability and show that this model only makes sense if it fits into a broader, user-centred approach to explore existing, potential, and desired practices. We illustrate the implementation of the model and its ability to represent usage cases of a home care structure digitizing the care record.

Keywords MDSEA interoperability · Enterprise modelling · Usage model

1 Introduction

Health organizations are involved for years in a global strategy for improving their performances and the quality of their business processes, due to the evolution of regulation as well as to external pressure. These strategies have an important impact on the processes and on the day-to-day work. A main characteristic of the health activities relies on the fact that business processes require a large set of stakeholders to be achieved, so that business processes are intrinsically collaborative processes. Interoperability is then a key point for structuring and improving health processes.

C. Merlo (✉) · V. Pilnière
ESTIA Institute of Technology, University Bordeaux, Technopole Izarbel, 64210 Bidart, France
e-mail: c.merlo@estia.fr

K. Borgiel
ISIS, INU Champollion, 81104 Castres, France

By the way, we consider that it is necessary to study interoperability from multi-organizations modelling to IT systems modelling, with a focus on health actors' activities.

In this paper, we focus on "home care" context, as defined by Bricon-Souf et al. [1], where the main organization coordinates several stakeholders, even from other organizations, for implementing the adequate collaborative process to the patient. The aim is to help the main organization to improve its business process by improving its IT system, especially by its digitizing. Our work is to analyse simultaneously the necessary changes for the organization and the introduction of new IT systems, then their mutual influence. According to Leonard-Barton [2], implementing a new IT system is a mutual and adaptative process between the organization and the technology that generates changes on several dimensions and at different levels of granularity. So, we must analyse and manage the impact of changes on collective and individual practices. Such a project focuses both on a business process reengineering approach and on a transformation of the IT system. It deeply impacts the organizational structure, and its success depends on its complex and evolutive context [3], by studying the interactions between actors and IT tools.

In the next section, we introduce the interoperability approach based on model-driven services engineering architecture (MDSEA) [4], then we apply it to a case study from health domain. In Sect. 3, we propose a "usage" model to characterize actors' practices to enrich MDSEA models. Then, we compare MDSEA and usage models and discuss of their respective interest.

2 MDSEA-Based Interoperability Approach

2.1 MDSEA Principles

In [5], we first proposed ISTA3 methodology, based on model-driven interoperability (MDI) principles, and derived from model-driven approach developed by Object Management Group [6], within the "INTEROP 2003–2007" Network of Excellence [7]. This approach was initially characterized in order to develop flexible interfaces between IT systems of both enterprises and to facilitate the use of a shared platform without leading to strong investments at the beginning of a collaboration. The main idea is to apply enterprise modelling, and especially GRAI approach, in order to represent and to analyse users' requirements and to transform these requirements at technical levels to specify IT solutions. Several levels are identified to propose a global architecture:

- The business service models (BSM) enable to collect needs and characteristics of the enterprise. GRAI models are used at this level e.g. extended actigrams and GRAI grids. The aim is to identify interoperability nodes from process and decision points of view that will be used to define future services supporting specific collaboration.

- The technology-independent models (TIM) aim at describing more deeply the collaborative processes through the IT system point of view. It can represent global specifications of the future information flows, based on the required collaborations (functions) identified at BSM level. The TIM models are derived from BSM models using mainly BPMN 2.0 formalism and model transformation tools.
- The technology-specific models (TSM) represent implementation models associated to the selected solutions (technologies or tools). They may correspond to detailed technical specifications (e.g. UML) before programming activities.

Main interest of this approach deals with the complementarity of the different levels that allows to understand the collaboration context and each partner's expectations at BSM level. Then, it ensures a continuum from this global vision of collaboration to the technical services that must be developed and orchestrated to support this collaboration.

2.2   Case Study for "Home Care" Organization

The health structure (Bayonne Health Service BHS), which is the "industrial" partner of our work, provides comprehensive home care services. It manages about 500 patients per day, and its activities are spread out on both hospitalization at home (HAD) for 20% and nursing at home (SIAD) for 80%. It employs more than 250 employees: medical, paramedical, administrative, etc., spread over more than 20 different professions. Despite this, to provide the patient with all necessary services, BHS relies on a large network of establishments and professionals in its territory, such as hospital and general practitioners, pharmacists, laboratories, and several liberal professions. It therefore offers global support through complex, multiple, geographically distributed business processes and activities. This characterizes a collaborative and distributed context, where the patient is at the heart of the concerns of a multitude of actors. The quality of care provided is ensured by a good level of coordination between the interventions, mostly asynchronous, of all actors. This coordination is facilitated using shared tools, such as the patient's record in paper format, located at patient's bedside.

In this context, the challenge of coordination is not only to manage the planning for home visits and for hospital and city trips, but also and above all to ensure continuity and consistency between the different care activities related to a given patient, despite the asynchronous nature of most interventions, and the random presence of persons performing the necessary functions with the patient. The patient's care record, located at home, is a crucial tool for tracking and reporting information between stakeholders. BHS is intended to replace the "paper" care record by a digital tool. It represents an opportunity to improve the way to carry out care activities. Expected objectives are to reduce information retrieval; to facilitate remote access to information; to improve the readability and completeness of information, and thus decision-making; finally, to ensure the "Electronic Prescription for the Care

Plan". For this project, the pilot team was composed of the doctor-director of the structure, the financial manager, and the research supervisor. The operational team was composed of a Ph.D. student, her two supervisors, the financial manager who oversees the project, and several executive nurses. The project have been composed of several steps inspired from GRAI integrated method [8, 7]: interviews and analysis of existing processes (AS IS situation); then AS IS modelling; diagnostic; design of the TO BE system including business processes improvements, interoperability study; finally elaboration of requirements for the IT solutions; experiments of the chosen IT tool based on existing and expected practices; iterations for improving the tool and experiments of the different versions of the tool.

2.3   Examples of MDSEA Models

Our aim is to improve the modelling of actors' practices, and we present hereby models from BSM and TIM levels describing business processes and information flows.

- **BSM level: modelling of the business processes**

During the AS IS steps, different processes of the structure have been modelled, firstly with a global vision to be able to correlate the different processes that take place. Main business processes of BHS are: managing the preliminary care process before accepting a patient; the care process itself; then, the post-care process. Several support processes have been identified. Secondly, a more detailed modelling was achieved for the pre-care, care, and post-care processes. The care process itself is decomposed into a coordination process and several sub-processes depending on the type of activity that has to be done at patient's home. The detailed models allow to identify and characterize collaborative activities. Interoperability problems are different for each of these collaborative activities' types and depend also on the actors work situation: is it an employee of BHS, an actor in a liberal profession, or an employee of another structure? What are their own business processes? What IT tools are they using? Fig. 1 shows the detailed modelling of the pre-care process. Two actors are BHS employees: the secretary and the nurse executive. The patient is considered as an external structure as well as the doctor which is often the usual doctor of the patient.

During TO BE phase, same models are achieved depending on the proposed IT and interoperability solutions. Information system architecture of BHS is modelled.

- **TIM level: modelling of the information flows**

Firstly, detailed BSM models are transformed in BPMN, such as in Fig. 2. Then, these models are detailed by introducing the different elements of the information system architecture and by developing the flows between the collaborative activities. Here, the pre-care process identifies a collaborative activity occurring at patient's home with the nurse executive, the patient, and the doctor. The aim is to analyse

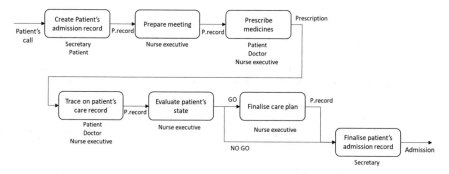

Fig. 1 Example of collaborative process models at BSM level, detailed view

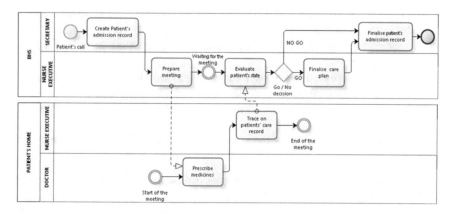

Fig. 2 Detailed collaborative process model at PIM level

patient's health situation, to define the possible care plan, and to make the medicine prescription for implementing the care plan. This collaborative situation will be reused later on.

- **TSM level: requirements and evaluation of the solutions**

At this last level, a specific software company has been chosen, and as a startup, a partnership has been established: BHS defined requirements, and the startup intended to implement the requirements in a generic way. Implementation allowed to make the IT solution evolve, both on functional aspects and on interoperable capabilities.

2.4 Limits of the Approach

The study of the information system of the different stakeholders (BHS and external actors/structures) coupled with the detailed BPMN models allows to identify and characterize requirements for solving the problems identified during the collaborative activities. When the startup has been able to supply a first version of the solutions, several problems appear because IT programmers did not know the future users' practices. So, users must modify their practices instead of using a tool adapted to their practices. Moreover, in the health domain, it is not possible to reduce all users to only one type of user. Considering previous situation at patient's home, the collaboration activity has been modelled as a unique activity. But in the real world, for 50 patients that we observed, nearly 50 different ways of achieving this activity have been identified such as:

- The patients' home has an access to internet, so the nurse executive is able to connect to a distant application for connecting to patients' record and to input new data.
- The patients' home has no access to internet, so the nurse executive must use its smartphone to connect to the distant application.
- "No internet and no GSM" is another situation, how to store some data for the future?
- The doctor prescribes using a paper sheet: how to introduce it in the application?
- The doctor has its own software to prescribe with its smartphone.
- The doctor accepts to prescribe with the distant application of BHS but wants a copy sent to its own IT tool.
- Or, the doctor wants to make the meeting at his own office and not at patient's home.

These short examples generate several types of solutions that must be implemented for the health structure, otherwise the cost of this IT project will override the expected benefits when transforming the paper-based records to digital records. To conclude on the limits, both GRAI models at BSM level and BPMN models at TIM level must be much more detailed to be able to trace real work. In the next section, we propose a model to observe, analyse, and characterize the practices of the individuals, with the objective of improving processes definition and requirements flexibility.

3   Proposal of a Usage-Centred Approach

In order to study the impacts between organization and digital tools, we focus on professional practices. From our point of view, the concept of practice emphasizes the man in a working situation, while the concept of usage emphasizes the object used by a human. When we talk about the positive or negative impacts that the digital tools can have on practices, we are interested in changes that facilitate or do not facilitate these practices, i.e. the fulfilment of the work accomplished. The fact that we consider also the concept of usage leads us to focus on the concept of the variability of usages of an IT tool and embodies the bidirectional link between the tool and the activity it supports. Several dimensions can be considered for describing usage [9, 8]; there is a link between individuals and a group of individuals when using a system [10, 9];

and the usages of a tool are built over time [11, 10]. The project of replacing one digital tool with another is to achieve a change "from an old way of doing things to a new way of doing things" [12, 11] and leads to organizational change. Moreover, as [10, 9] points out:

- different people in the collective may use different functionalities of the (technological) system;
- different people in the collective can use the system for different tasks;
- and more generally, different people in the collective can use different features for different tasks.

Our first proposal is to define a model of usage able to trace the multiplicity of linkages between the organization and the technology. The definition of the user's model is based on the identification of a set of qualitative variables that characterize the usage of an IT tool. In order to define the usage variables that characterize the conceptual model of usage, we use the 5W2H method of analysis (QQQOQCCP in French): **Who? What? Where? When? How? How many? Why?** This method serves as a general guide for gathering comprehensive information on a situation [13, 12]. Based on our analysis of the bibliography and the industrial context, we first identify four variables that we associate with the following descriptors:

- **Who**? Individual (the one who acts)
- **What**? Activity (what is done)
- **How**? IT tool (object used during the activity)
- **Where**? Location (of the activity)

We refer to this first group of four variables as the characterization of a "**usage case**". We complete this description with a fifth variable, **When**, which represents the temporal dimension and allows to describe the sequence of several activities [14, 13]. We refer to this set of activities as a "**usage scenario**". This scenario is an extension of the usage case, either to place it into a business process or to further detail it.

3.1 Application to Health Case Study

These variables are useful to describe the results of an observation of actors doing a collaborative activity, but we quickly understand that we must choose a more graphical formalism to be able to communicate and exchange with the actors. After some meetings using traditional engineering process models (e.g. actigrams, BPMN), we had to define a graphical formalism much easier to manipulate by all the actors of the project, actors without engineering background. In Fig. 3, the different graphical objects are represented. We describe the activity "prescribe medicines". The situation is described on the left, showing the location, the individuals involved, and the tools. Several situations can be identified for each BPMN activity that we want to focus on.

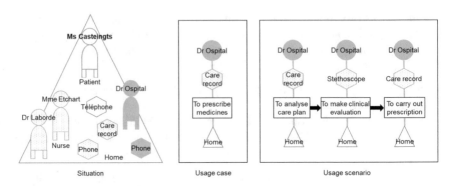

Fig. 3 Usage model: situation, usage case, and usage scenario

In the middle, the usage case is described: "Dr Ospital" (Who) uses the "care record" (How) to help him "prescribe medicines" (What) at patient's home (Where). For each situation, one or more usage case can be defined. On the right, a usage scenario is defined to detail the tasks sequence of the usage case: Dr Ospital analyses the care plan using the care record, then makes a clinical evaluation of the patient, with e.g. a stethoscope, and finally carries out the prescription that will be stored in the same care record. Of course, several usage scenarios are possible for one usage case.

For example, Fig. 4 shows an alternative usage case where the care record, in a paper format, is replaced by an IT tool, and the location is the doctor's office. The five variables identified for the usage model allow defining a usage case and the resulting usage scenario(s).

Together, they allow to understand a complete usage and how it is done. They also allow characterization and study of the diversity of usages. Our observation field demonstrates that from a first usage case, other associated usage cases may exist with different values for one (or more) variable(s), or that several usage scenarios

Fig. 4 Usage model: diversity of usage case "To prescribe medicines"

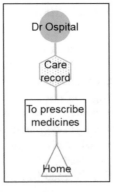
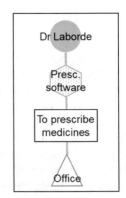

Usage case: alternative 1   Usage case: alternative 2

may correspond to the same initial usage case. Finally, many industrial engineering works on the usages or the modelling of organizations and business processes show the importance of distinguishing different levels of granularity when describing an activity.

So, we defined four levels of granularity for each variable describing a usage case:

- **Who**? Individual → Function/Role → Group → Business Collective.
- **What**? Action/Operation → Activity → Business Process → Organizational Function.
- **How**? Information/Data → Component → SI Tool → SI Business.
- **Where**? Position → Part → Site → Business Space.

These different levels allow a more precise description of a general usage case, or a more detailed description of a usage scenario than the usage case to which it refers, without all variables being at the same level of granularity.

We discuss now of the interest of this model, on the way we used it for improving the generic solutions that were proposed during the MDSEA approach.

3.2   Enrichment of MDSEA Models and Discussion

Changes in the information system cover the generation, implementation, and adoption of new elements in the organization's social and technical subsystems that store, transfer, manipulate, process, and use information [14, 15]. The use of the usage model is part of a general approach and methodology centred on actors. Indeed, in this approach, man is seen as an actor when achieving his work that manages the requirements of production, quality, deadlines, and also the hazards, dysfunction, failures, fatigue, relations with colleagues, with the management.

3.2.1   Application of the Usage Model

This focus on the man at work enabled us to identify work situations encountered by professionals throughout the patient's care. The various processes were modelled and used as a basis for discussion to integrate usage modelling.

The usage model was first used for understanding existing situations of work identified when modelling collaborative activities with GRAI actigrams, then BPMN diagrams. As we have shown in the previous examples, the usage model allows us to detail much better the collaborative activities and to really understand the usage cases, then all the variability of such usage cases through the usage scenarios. As a consequence, we were able to identify several "users' profiles" where traditional approach identified only one profile. Then, we apply the usage model to build the experimentation of the interoperable IT solutions: instead of generating one sequence of tasks for one type of users, we were able to generate several different sequences according to the different scenarios and the different profiles previously identified. Moreover, the sequences were not built on the "existing scenarios" but on the "expected scenarios" based on the new IT solutions. Then finally, we reused the usage model a third time to validate the usage scenarios to be implemented when the experimentation generated "good" results, i.e. the different types of users validate the tool, and the managers validate the resulting added value on the business process.

3.2.2   Impact on the Initial MDSEA Approach

Therefore, we conclude from these three-step applications that usage model is helpful:

- In the AS IS phase for an enrichment of the TIM level and for a better communication support with actors that have no engineering background;
- In the TO BE phase for more detailed evaluations of the interoperable IT solutions, made at the TSM level, by exploring more variability from the solutions and by allowing technical decisions to be linked to their impact on human and process decisions;
- In the TO BE phase at the BSM/TIM levels by taking decisions for modelling the TO BE GRAI actigrams/BPMN diagrams, but after the evaluations at TSM level, and before the implementation of the IT solutions again at TIM level.

Due to the identification of five variables for usage modelling "Who, What, How, Where, When", we consider that the usage cases or the usage scenarios could be integrated into GRAI actigrams or BPMN diagrams as activities or sequence of activities (what, when), with main characteristics: resources (who, how), input/output information (how). But, it is not always relevant due to the different levels of granularity.

3.2.3   Limits of the Usage Model

Implementing such model was facilitated by the participation of a Ph.D. student all along the project. But, it is time-consuming for making numerous observations and being able to identify all the different types of actors and scenarios. Moreover, several iterations of experimentations and modifications of the IT solutions were necessary. Even if final TO BE decisions were taken with deeper justifications, this takes also a long time to analyse all the different types of situations. Three years were necessary from initial interviews to the deployment of the IT solutions. Similar MDSEA approach in previous industrial case studies with similar structure size took less than two years.

3.2.4   Application of the Usage Model

The proposed usage model has been applied in the health domain, which is characterized by processes strongly collaborative where coordination between actors is a great challenge. We consider then that the proposed approach should be available in other domains presenting similar characteristics, such as product/system development or supply chain processes. Further work is expected for demonstrating it.

4   Conclusion

In this article, we have been interested in taking professional practices into account in technological change. To this end, we have developed a model called "usage model" which we have illustrated with a case study in the health domain, precisely home care. We stress the importance of integrating this model centred on man at work into a re-engineering approach such as MDSEA (GRAI) approach that allows collaborative work. Without this precaution, the model loses its meaning and therefore its interest, and it becomes only a tool that cannot be used and manipulated properly and which will not contribute to the desired result, namely a positive change in professional practices. We show that such a model allows enriching BSM and TSM levels process modelling by adding more details on the real individuals' tasks. The several variants of tasks sequences associated to one situation and to one profile allow to better evaluate interoperable IT solutions and to take decisions deeply justified for the future IT solutions and the future business processes, from high-level ones to very detailed ones. Future work will focus on the definition of an integrated methodology based both on MDSEA approach and usage model implementation. We thank BHS and all actors involved in this project for their motivation and their patience before its positive ending.

References

1. Bricon-Souf, N., Anceaux, F., Bennani, N., Dufresne, E., & Watbled, L. (2005). A distributed coordination platform for home care: Analysis, framework and prototype. *International Journal of Medical Informatics, 74*, 809–825.
2. Leonard-Barton, D. (1988). Implementation as mutual adaptation of technology and organization. *Research Policy, 17*(5), 251–267.
3. Lynn Crawford, J. P. (2004). Hard and soft projects: A framework for analysis. *International Journal of Project Management, 22*.
4. Chen, D. (2014). A set of templates for MDSEA. In *Workshop on model driven services engineering architecture (MDSEA): A result of MSEE Project.* Enschede.
5. Merlo C., Vicien G., & Ducq, Y. (2014). Interoperability modelling methodology for product design organisations. *International Journal of Production Research, 52*(15), 100–120, https://doi.org/10.1080/00207543.2013.774484.

6. Object Management Group, "MDA Guide Version 1.0" document number: omg/2003–05, http://www.omg.org/mda/mda_files/MDA_Guide_Version1-0.pdf.
7. INTEROP NoE: Deliverable DTG 2.3—Report on Model Driven Interoperability (2007).
8. Doumeingts, G., & Ducq, Y. (2001). Enterprise modelling techniques to improve efficiency of enterprises. *International Journal of Production Planning and Control, 12*(2), 146–163.
9. Ammenwerth, E., Iller, C., & Mahler, C. (2006). IT-adoption and the interaction of task, technology and individuals: a fit framework and a case study. *BMC Medical Informatics and Decision Making, 6*(3).
10. Burton-Jones, A., & Gallivan, M. J. (2007). Toward a deeper understanding of system usage in organizations: A multilevel perspective. *MIS Quarterly, 31*(4), 657–679.
11. Fidock, J., & Carroll, J. (2012). Theorizing about the life cycle of IT Use: An appropriation perspective. In D. N. Hart & S. D. Gregor (Eds.), *Information systems foundations: Theory building in information systems* (pp. 79–112). ANU E Press.
12. Lorenzi, N. M., & Riley, R. T. (2003). *International Journal of Medical Informatics, 69*, 197–203.
13. Crampton, N. J. (1998).Preventing Waste at the Source. CRC Press. ISBN 10: 1566703174.
14. Borgiel, K., Merlo, C., & Minel, S. (2015). A multi-level activity analysis for home healthcare ICT tool redesign. In C. Weber, S. Husung, G. Cascini, M. Cantamessa, D. Marjanovic, & M. Bordegoni (Eds.), *Proceedings of the 20th International Conference on Engineering Design (ICED 2015)* (Vol. 1, pp. 475–484). Milan.
15. Lyytinen, K., & Newman, M. (2008). Explaining information systems change: A punctuated socio-technical change model. *European Journal of Information Systems, 17*, 589–613.

Combining Reference Models for Eliciting Requirements in Industry 4.0 Projects: A Demonstration Case

Nuno Santos[ID]**, Jaime Pereira**[ID]**, Francisco Morais**[ID]**, João P. Mendonça**[ID]**, and Ricardo J. Machado**[ID]

Abstract The industrial domain has faced an increase of complexity, mainly due to recent technological evolutions—from sensors, connectivity, platforms, etc. Main consortiums in this domain have proposed their reference models to ease development of Industry 4.0 (I4.0) or Industrial Internet of Things (IIoT); however, companies still struggle to design their architectures. This paper proposes how reference models can be combined for early design decisions, which impact how requirements are elicited. By combining not only I4.0/IIoT references such as IIRA and RAMI4.0, this paper describes the adoption of references at the cloud and at the edge level. The approach then uses model-based development in order to define scenarios, requirements, and architecture components and deployment. The approach is described using an IIoT research project as a demonstration case.

Keywords IIoT · I4.0 · Requirements elicitation · Reference models · RAMI4.0 · IIRA · NIST-CCRA · OpenFog RA · Architecture design

1 Introduction

To face the demand of Industrial Internet of Things (IIoT) projects, a plethora of reference architectures have been developed in several domains [1]. Examples such as the Industrial Internet Reference Architecture (IIRA) [2], *Industrie 4.0* Reference Architecture Model (RAMI 4.0) [3], and NIST Smart Manufacturing (NIST SM) [4] provide standardization for developing industrial architectures. On the technology

This work was developed within the UH4SP: Unified Hub 4 Smart Plants (Project ID 017871), under Portuguese National Grants Program for R&D projects (P2020–SI IDT), COMPETE: POCI-01-0145-FEDER-007043, and by FCT—Fundação para a Ciência e Tecnologia within the R&D Units Project Scope: UIDB/00319/2020

N. Santos (✉) · J. Pereira · F. Morais · J. P. Mendonça · R. J. Machado
CCG/ZGDV Institute, Guimarães, Portugal
e-mail: nuno.a.santos@algoritmi.uminho.pt

N. Santos · J. Pereira · F. Morais · R. J. Machado
School of Engineering, ALGORITMI Center, Minho University, Guimarães, Portugal

adoption side, another example is the NIST Cloud Computing Reference Architecture (NIST-CCRA) [5], which is a standard within the deployment of cloud solutions, which are a crucial topic in IIoT. While IIRA claims to be applicable regardless of the industrial domain, RAMI 4.0 classifies itself as a reference for the manufacturing domain [6]. RAMI 4.0 and IIRA are not exclusive in terms of their adoption, as there are studies regarding implementing testbeds semantically interoperable from a functional point of view [7, 8].

Adopting these reference architectures is just the starting point in developing IIoT systems [9, 10], whereas design and development require other approaches. One possible approach is using model-driven architectures/development, which has previous experiences in development of technology that is relevant for these environment, such as the case of cloud computing solutions [11] and fog computing architectures [12]. Additionally, literature about IIoT architectures focus in the mentioned reference architectures, although some present specific and concrete use cases where IIoT technologies are being applied, like for instance asset efficiency testbeds. Even though such literature provides insights regarding architecture implementations, they are insufficient when implementing IIoT projects.

In order to kick off any IIoT projects, it is advisable to have clear views in both business (organizational) and technology (IIoT) constraints, with a harmonized view of the organizational processes and the roles of systems within them. Reference models encompass both views properly aligned, but a view on the organizational context is also required. This paper proposes combining architecture reference models, namely adopting their classification of concerns, with elicitation techniques for the specific business processes. With this approach, we aim to provide a more structured and representative view of the most fundamental design decisions. This paper uses a research project for the IIoT domain, called Unified Hub for Smart Plants (UH4SP), as a demonstration case of using this approach along the analysis, design, implementation and deployment phases of an IIoT project.

This paper is structured as follows: Sect. 2 introduces the UH4SP project as a running example for the remaining sections; Sect. 3 presents using reference models for separation of concerns; Sect. 4 uses a scenario-based requirements elicitation; Sect. 5 describes architectural design levels; and Sect. 6 presents the conclusions.

2 Running Example: The Unified Hub for Smart Plants (UH4SP)

The UH4SP project envisioned a centralized architecture for integrating data from distributed industrial unit plants. The objective was to centralized business and operational data and use it as a common ground for a set of new services to be developed into the entire supply chain (plants, suppliers, forwarders, and clients). It also encompasses new services toward a corporate management of production, where a group of industrial plants have a global view of their plants' performance and process efficiency. Data would originate from ERP systems (like SAP), MES and weighting systems developed by *Cachapuz Bilanciai Group*, located in Braga, Portugal.

The project aimed validating such assumptions within a proof of concept performed in an ecosystem of industrial unit plants, where process monitoring would focus in the arrival and exit of trucks in the industrial unit plants facilities, as well as to control load and unload activities. Additionally, such monitoring and management would have to be supported by communication of the platform with the plant's ERP and the industrial hardware.

3 Separation of Concerns

As the industrial environment is increasing in complexity and composed by a plethora of systems, reference architecture models are a way of organizing those systems in a standardized structure. Using the reference architecture's layers allows to define proper ways to integrate and interoperate with each of the involved systems. Since some reference models' layers can be mapped and intertwined between each other, it is possible to define a model that combines a set of reference models as best suited for the organization. For the sake of this research, we adopted the ISA 95 "Enterprise—Control System Integration," IIRA and RAMI4.0 models. A mapping between layers of these models is depicted in Table 1. Such mapping is based in previous research from [8] and [13]. The three reference models are now briefly introduced for the sake of understanding the layer mapping.

The ISA-95 model is hierarchy-based in five business process levels. Level 4 is the highest in the hierarchy and relates to "business planning and logistics." As the name implies, these systems are associated the business management, for instance, enterprise resource planning (ERP), among others. Level 3 relates to "manufacturing operations management." These systems relate to manufacturing management, like manufacturing execution systems (MES), among others. Level 2 relates to "monitoring, supervisory control, and automated control of the production process." These systems are concerned about management of operations, for instance, supervisory control and data acquisition (SCADA) and human–machine interface (HMI) systems. Level 1 relates to "sensing and manipulating the production process," associated with direct interaction with the operations, for instance, programmable logic controller (PLC) systems. Level 0 is the lowest in the hierarchy and relates to "the actual production process," which refer to machines, devices, and the resources.

The IIRA model uses similar splitting of concerns, which they define as viewpoints. This model includes the following four viewpoints: Business, usage, functional, and implementation. Due to the broad scope of the model, for the purpose of this analysis, we focus in the functional viewpoint, because it is related to the involved systems and includes additional viewpoints: "Business," "operations," "information," "application," "control" (sense or actuation), and "physical systems."

Table 1 Layer mapping between ISA-95, IIRA, and RAMI4.0

ISA-95	IIRA	RAMI4.0	Examples
Business planning and logistics (Level 4)	Business	Business	Enterprise resource planning (ERP)
Manufacturing operations management (Level 3)	Operations, application	Functional	Manufacturing execution systems (MES)
	information	Information	
N/A	crosscutting functions (connectivity, distributedData management)	communication	Gateways
Monitoring, supervisorycontrol and automated control oftheproduction process (Level 2)	Control	Integration	Supervisory controland data acquisition (SCADA), human–machine interface (HMI)
Sensingand manipulating the production process (Level 1)			Programmable logic controller (PLC)
actualProduction process (Level 0)	Physical systems	Asset	Machines, devices, and the resources

Finally, RAMI 4.0 includes layers that are classified by vertical systems, facilities, and products lifecycle, each one referring to a given axis in the model. These axes then reflect different hierarchy levels. Just like the IIRA, for the purpose of this analysis, we only focus in the vertical systems axis: "Business," "functional," "information," "communication," "integration," and "asset."

It is proposed that, by defining software components within these layers, one may use such reference components as basis for elicitation of functionalities. Moreover, other main concern derived from these layers is the need for addressing integration, communication, or interoperability requirements whenever there are flows between layers. For instance, systems like ERPs may have tendency to use APIs, SOAP or HTTP protocols. SCADAs or PLCs systems may use OPC-UA, MQTT, or AMQP. It also depends on the layer the other system involved in the communication is positioned. The reference architectures, as the ones discussed in this section, define best practices in communications between layers and thus are core sources for this elicitation. It shall be referred that at this point it is targeted "how" information will flow. However, before that it must be defined "what" information flows, which is discussed in the next section. For the separation of concerns discussed in this section, let us refer to the running example. The UH4SP project started by this precise analysis. The consortium identified the industrial reference models suitable for the project to follow based in the project's objectives. Overall, the consortium identified IIRA and

RAMI4.0 as the relevant references for the project. Just like in [8], by combining IIRA and RAMI4.0, an architecture with three layers is proposed using the same layers: the enterprise tier, the platform tier and the edge tier. This approach led to a separation of the concerns between each layer. For the enterprise tier, the project included applications for end-users. Additionally, this tier also includes systems that deal with operational data from each industrial unit. These systems are typically operating at the shop floor level. Some interfaces would have to be developed for enabling operational information to be available. The platform tier relates to the cloud services to be developed. These services will be consumed by the applications from the enterprise tier, but also invoking edge services in the edge tier as well. Each edge was going to be deployed as responsible for each industrial unit, gathering the data from the operational systems.

Each layer is able for further reference adoptions. For example, it was analyzed how NIST-CCRA could be used as basis for designing the architecture for the cloud tier and OpenFog Reference Architecture for designing the edge tier.

4 Scenarios Toward Software Requirements

Now that the technology decisions are structured; this section now focuses in understanding the business side. An initial understanding of the domain is required. For that, the identification of business needs started by comprehending the involved actors and how they will need to interact with the new solution. Afterward, functionalities are elicited using UML models.

The requirements elicitation started by listing a set of stakeholder expectations, which promoted the discussion of scenarios (Fig. 1). Afterward, the requirements analysis included gathering the scenarios and elicit the functionalities, in form of a UML Use Case diagram. The use case model of the UH4SP project (Fig. 2) was composed by 37 use cases, elicited by combining functionalities from reference architectures and the scenarios.

Use cases from *{UC.2} Configure cloud service* (Fig. 2a) were derived from NIST-CCRA "provisioning / configuration" layer, mainly allowing to include in our model monitoring, metering, deployment models and service agreements that the cloud platform must include. Use case *{UC.4} Manage cloud security and privacy* (Fig. 2b) were derived from NIST-CCRA "security" layer. The elicited functionalities aim at managing backups and monitoring activities. Other layers from NIST-CCRA were included in the use case model, like "business support" (like users profile, costumers accounts and their licenses), "data portability" (also addressed in use cases *{UC.3.2} Synchronize data* and *{U.C.7.2.4} Integrate local information systems data*) or "privacy" (for users accounts and profiles configurations). Use case *{UC.3} Manage cloud interoperability and portability* (Fig. 2c) were derived from OpenFog RA, referring to synchronizing data from local information systems, and the required data management, at the edge layer.

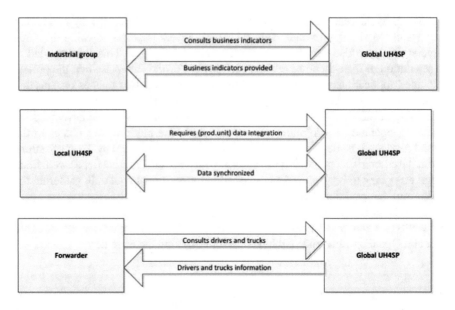

Fig. 1 Scenarios elicited

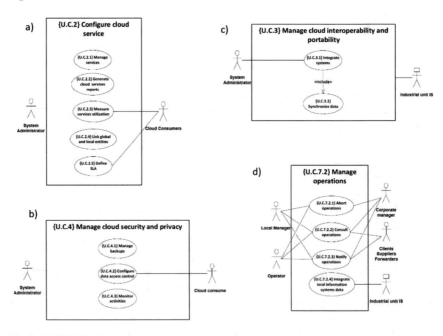

Fig. 2 UH4SP first-level use cases

Use cases from *{UC.7}* all refer to performing business processes in the scenarios from Fig. 2., where we depict a sub-set of them in Fig. 2d—*{UC.7.2} Manage operations*, features related to operations management, namely: to abort a given logistic operation, to consult operations that where information comes from *{U.C.7.2.4} Integrate local information systems data* and synchronized at *{UC.3.2} Synchronize data*, and to perform notifications about industrial unit logistic operations.

5 References Used for Architectural Design

Section 3 discusses how reference models could be adopted for an initial separation of concerns. Such separation is useful, for instance, if a different team may be responsible for developing one specific concern. However, reference models can be continuously present in the project development, and in this section, we will discuss their use in architectural design.

Design is a task which takes in consideration the domain and business needs (cf. Section 4) and design decisions from technology constraints in order to meet quality requirements. After the elicitation of the business needs, architecture design should address how such needs can be technologically supported, in form of software components. It is more frequent that design focus in supporting those business needs than following any reference model. However, reference models, and the separation of concerns that are derived by them, allow to understand, e.g., communication issues between layers.

Inside each layer, other references provide useful guidance as well. It has already been mentioned the use of NIST-CCRA for cloud implementation, as well as the OpenFog Reference Architecture (OpenFog RA) for fog architectures. Additionally, other reference models like ETSI-MEC (Mobile Edge Computing), OPC-UA, Open Connectivity Foundation (OCF), OpenNFV, among others, may also be adopted. It is also worth referring that these models are complementary to IIRA and OpenFog.

The UH4SP project architecture was designed with five major packages: *Configurations*; *Monitoring* (Fig. 3); *Business management*; *Integration*; and *Fog data*. This

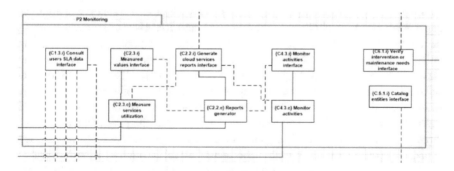

Fig. 3 *Monitoring* package from the UH4SP architecture

structure heavily relies in the OpenFog RA, with separate packages concerning cloud, fog, and edge. Additionally, for the cloud services, NIST-CCRA was used.

The architecture was design with 77 architectural components, reflecting the business needs identified in the 37 use cases from the previous model, in form of UML Components. The logic behind the components architecting was to use design decisions from model-view-controller (MVC).

Finally, the deployment of the functionalities was also addressed. It is natural that deployment architecture also reflects the separation into the layers. At this stage, adopting reference architectures is actually more reflected in the resulting architecture, where the deployment architecture depicts the deployment location of the applications. The deployment architecture for the UH4SP project used deployment decisions in three layers just like IIRA, as systems and services are deployed in an enterprise, platform and edge tier. One layer where business applications interface directly with human actors, and within operational systems where information about production of a local industrial unit is generated. The edge layer is located—as the name implies—at the edges, for each industrial unit. For the cloud services layer, services for supporting the business applications by means of a microservices-oriented architecture.

6   Conclusions and Future Work

The complexity of IIoT projects, due to the large number of involved industrial systems, their heterogeneity, as well as the emergence of new technologies at a rapid pace, has led to companies often striving to properly elicit functionalities and design solutions. References and standards in this domain are still recent and immature, so companies are unable to easily design solutions based upon them.

The proposed combination of separation from concerns within requirements elicitation is the starting point for IIoT projects, namely contributing for the typical domain analysis tasks that are performed in early stages of the project.

The reference models and separation of concerns were used in this paper as basis for defining the entities that interact for a set of scenarios. Reference models allow aligning organizational and technology concerns. Through a running example, this paper described a combined adoption of IIRA and RAMI4.0 at an early stage for defining separation of concerns. An edge layer, aimed to support groups of industrial unit plant's data, was specified using OpenFog RA. For specifying the several cloud services, as well as the platform's management, the NIST-CCRA was used. Such separation of concerns allowed a view of the distribution of systems and actors, where scenarios were elicited so information flows could be identified. A model-based approach, like UML Use Cases, allowed to define the project's requirements through such scenarios. Models then evolve toward solution's architecture design, supported from domain and requirements analysis level and ending at the deployment level. It also allowed identifying communication needs, namely between the layers.

This research has still some points to be addressed in the future. There is a lot to improve the support for design decisions, namely relations between the deployment design and the reference architectures (besides hierarchies). Additionally, reference models are not described at the same abstraction level, so the evolution of models within our projects requires an adequate adoption of those references according to their abstraction level, which will be targeted in future research.

References

1. Weyrich, M., & Ebert, C. (2016). Reference architectures for the internet of things. *IEEE Software*. https://doi.org/10.1109/MS.2016.20
2. IIC. (2017). *The industrial internet reference architecture (IIRA) v1.7*.
3. Hankel, M., & Rexroth, B. (2015). *The reference architectural model industrie 4.0 (RAMI 4.0)*. ZVEI.
4. Lu, Y., Morris, K. C., & Frechette, S. (2016). *Current standards landscape for smart manufacturing systems*. NIST.IR.8107.
5. Bohn, R. B., Messina, J., Fang, L., et al. (2011). NIST cloud computing reference architecture. In *IEEE World Congress on Services (SERVICES)* (pp. 594–596).
6. Lin, S. -W., Mellor, S., Munz, H., & Barnstedt, E. *Architecture Alignment and Interoperability An Industrial Internet Consortium and Plattform Industrie 4.0 Joint Whitepaper Contributors 1*.
7. Lu, Y., Morris, K. C., & Frechette, S. (2015). Standards landscape and directions for smart manufacturing systems. In *2015 IEEE International Conference on Automation Science and Engineering (CASE)* (pp. 998–1005). IEEE.
8. Infosys. (2016). *Interoperability between IIC Architecture & Industry 4.0 Reference Architecture for Industrial Assets*.
9. Monteiro, P., Carvalho, M., & Morais, F., et al. (2018). Adoption of architecture reference models for industrial information management systems. In *9th IEEE TEMS International Conference on Intelligent Systems (IS2018)*. IEEE.
10. Mell, P., & Grance, T. (2009). *The NIST Definition of Cloud Computing*.
11. Santos, N., Ferreira, N., & Machado, R. J. (2017). *Transition from information systems to service- oriented logical architectures: Formalizing steps and rules with QVT*.
12. Santos, N., Rodrigues, H., Pereira, J., et al. (2018). Specifying software services for fog computing architectures using recursive model transformations. *Fog computing: Concepts, frameworks and technologies* (1st ed., pp. 153–181). Springer.
13. Kemppainen, P. (2016). Pharma industrial internet: A reference modelbased on 5G Public private partnership infrastructure, industrial internet consortium reference architecture and Pharma industry standards. *Nordic and Baltic Journal of Information and Communications Technologies*, 141–162.

Data and Knowledge Modeling

Modeling and Sharing Knowledge in Expertise Processes

Serge Sonfack Sounchio, Laurent Geneste, and Bernard Kamsu Foguem

Abstract Expertise processes are exploratory and incremental, i.e., they are not defined a priori, but their structure evolves over the course of the expertise. These processes are omnipresent in companies but remain little studied. In this article, we propose to characterize these processes and to propose mechanisms to facilitate their realization. Accordingly, a general answer set to knowledge representation (Answer set programming (ASP)) and a World Wide Web Consortium (W3C) Recommendation (The Rule Interchange Format (RIF)) are used to improve knowledge sharing and interoperability in a cooperative expertise framework. Therefore, in this paper, we translate an Answer Set Program (ASP) solving approach modeling an expertise to a Rule Interchange Format using the Core Answer Set Programming Dialect (CASPD). We start by presenting how expertise processes are being carried out based on the NF X50–110 standard, followed by a presentation of Answer Set Programming and Rule Interchange Format. We end up by illustrating the translation of a car diagnosis expertise from ASP to CASPD.

Keywords Exploratory processes · Semantic web · Non-monotonic logic · Rule interchange format · Answer Set Programming

1 Introduction

For several years now, companies have been setting up structured and formalized processes. However, certain processes (or parts of these processes) do not respect these prior requirements for the formalization of a working framework. These are highly exploratory processes, which phases are not well known. In this context, we can cite, for example, the search for the causes of a problem in a company, the innovative design of a system or the analysis of a situation carried out by an insurance

S. Sonfack Sounchio (✉) · L. Geneste · B. K. Foguem
Laboratory of Production Engineering (LGP), EA 1905, ENIT, INPT, University of Toulouse, Toulouse, France
e-mail: serge.sonfack_sounchio@enit.fr

company. In this article, we are interested in the formalization of these expertise processes and the mechanisms that could be implemented to facilitate them [1].

An expert according to [2] is a person with trained in a specific field and capable of solving difficult problems based on what his had learned or acquired with experience. He brings his understanding of a problem at the request of another person, this is called expertise by Grundmann [3]. In other words, expertise is the application of the knowledge of an expert in a field in order to advise the applicant or help him in his decision-making.

Our focus is to show how one can use a language that support commonsense reasoning to carry out an expertise and share the rules or methods used in a human or systems readable format. To do so, we presented how expertise is carried out following NF X50–110 standard [4]; it is followed by a concise thought on Answer Set Programming and the W3C Rule Interchange Format (RIF). We round up with a use case from car hard to start diagnosis modeling with ASP and translated to a rule interchange format.

In order to carry out a good expertise, it is important to follow a well-structured process. For that reason, the **NF X50–110** [4] standard, was created. It clarifies steps or processes of expertise and how they are conducted. This defined process helps in:

- bringing transparency to the expertise process
- facilitating exchanges between actors involved in the expertise process
- justifying the results of the expertise

First, an expertise is both an exploratory and incremental process. Exploratory because at each level or step, one doesn't know the outcome or outcomes of the current step. It is also an incremental process because, to have an answer or a solution to a particular problem one need outcomes of the previous steps.

Looking at its structure, an expertise process can look like Fig. 1, based on the number of experts involved and the complexity of the domain or the problem to solve. For each step in the process, a goal is defined and at the end of the step, the achieved goal is used in the next step. It is also possible to have many steps going on at the same time.

Important point within a step of an expertise are:

- Exploration
  It provides information about the current question to answer. At this point, we find activities like surveys, measurements, research and studies of the question.
- Results interpretation
  This activity is to compute significant results from exploration activity. It goes from simple values to the use of statistical tools.

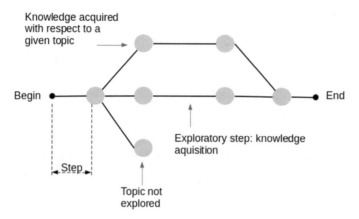

Fig. 1 Example of expertise process

2 Answer Set Programming (ASP)

Expertise processes, as described in the previous section, are of an exploratory nature. This exploration is mainly based on uncertain assumption formulation that may be confirmation/refutation. The process requires facts and reasoning that can, at a given state, support or disproof previous assumptions.

Looking at this important points of expertise process, one of the best reasoning paradigm that covers them, with the expressiveness of knowledge representation and reasoning is Answer Set Programming.

ASP is a programming language based on procedural programming, which consists of describing a problem and finding solution sets using a solver [5, 6]. We use this declarative paradigm for modeling and solving combinatorial search problems and for knowledge representation and reasoning [7, 8]. This language is suitable for solving optimization problems and derives its roots from logical programming, knowledge representation and non-monotonous reasoning. ASP is very closed to Boolean satisfiability problem (SAT) and constraint programming which are been used in inductive reasoning [9]. The non-monotonicity is achieved with a form of negation, called negation as failure or default reasoning [10], making ASP suitable for common-sens reasoning. This way of reasoning is not too far from human reasoning because while reasoning, it is possible to infer new knowledge from new information different from the current knowledge.

These characteristics make ASP an ideal language for expertise processes because expertise processes as described in the early section are based on exchanges between experts and users, and hence, these can yield incomplete information and be subject of contradictory assertions. Three important points make ASP different from Prolog:

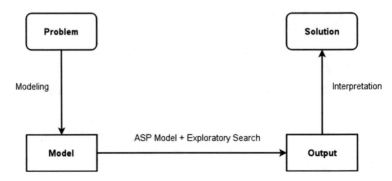

Fig. 2 ASP workflow

- One of the particularities of ASP over Prolog is the ability to learn non-deterministic concepts, using what is called choice rules and hard constraints in the case where we have a body, for example, to learn that a coin non-deterministically land on either heads or tails but never on both 1{heads, tails}1.
- The other big difference between ASP and Prolog is the possibility to extend inductive logic programming with the learning of preference models, using weak constraints.
- The search hypothesis in ASP is bottom-up, while Prolog uses the top-down. The bottom-up starts with the most specific clause and generalizes while keeping most negative ones out (Fig. 2).

2.1 Syntax

An ASP program is written as a pair $\{\sum, \Pi\}$, where $\sum$ is the alphabet and Π a finite set of **normal rules**, **constraints** and **choice rule**.

$\sum = < \mathbf{O}, \mathbf{F}, \mathbf{P}, \mathbf{V} >$ also called its signature, where $\mathbf{O}$ is a set of objects, $\mathbf{F}$ is a set of functions, $\mathbf{P}$ is a set of predicates, $\mathbf{V}$ is a set of variables.

The following elements are important for ASP programming [11]:

- Terms over $\sum$ are variables or constant objects
  if $t_1, \ldots, t_n$ are terms and $\mathbf{f}$ a function symbol of arity $\mathbf{n}$ then $\mathbf{f}(t_1, ..., t_n)$ is a term.
- $\mathbf{p}(t_1, ..., t_n)$ is an atom if: $t_1, ..., t_n$ are terms and $\mathbf{p}$ is a predicate.
- A literal is: an atom $\mathbf{p}(t_1, \ldots, t_n)$ or its negation $not\mathbf{p}(t_1, \ldots, t_n)$
- A ground atom is one with ground terms
- ASP rule has the following format:
  $h \leftarrow b_1, b_2, \ldots, b_n, notb_{n+1}, notb_{n+2}, \ldots, notb_m, n, m \in \mathbf{N}$
  h: is called the *head*
  $b_1, b_2, \ldots, b_n, notb_{n+1}, notb_{n+2}, \ldots, notb_m$ is called the *body*
  h, b_i are atoms.

- Constraint
  A constraint is a rule without a head (h)
  $\leftarrow b_1, b_2, \ldots, b_n, not b_{n+1}, not b_{n+2}, \ldots, not b_m.$
  Constraints represent thinks that are impossible to believe.
  Example: $\leftarrow$ p(a). " It is impossible to believe p(a) "
- Fact
  A fact is a rule without a body. $b_1, b_2, \ldots, b_n, not\ b_{n+1}, not\ b_{n+2}, \ldots, not\ b_m.$
  Facts are things that we believe (beliefs whether justified or not).
  Example: p(a). " Believe p(a)"
- Choice rule
  $k\{h_1, \ldots, h_m\}u \leftarrow b_1, b_2, \ldots, b_n, not\ b_{n+1}, not\ b_{n+2}, \ldots, not\ b_m$
  $l, n \in \mathbf{N}$. It allows to learn non-deterministic concepts, where k is the lower bound and u the upper bound of the cardinality of the alternative ways to form the stable models described by the rule.
- Save variable
  A variable V that occurs in a rule is said to be *save*, if V occurs in at least one positive literal in the body of that rule.

These elements are use to describe objects of a domain and relations between them. The meaning of a program defines the possible set of beliefs (answer set) [11].

2.2 Inference

Satisfiability: Given a program Π decides whether it has at least one answer set Brave Reasoning: Given a program Π and a ground literal Q, decides whether some answer set satisfies Q; we then say that Q is a brave consequence of Π. Cautious Reasoning: Given a program Π and a ground literal Q, decides whether all answer sets satisfy Q; we then say that Q is a cautious consequence of Π [12]. Answer sets are built by ASP solvers from the following principles:

1. Satisfy rules.
2. Do not believe in contradictions
3. Adhere to the **Rationality Principle**: "Believe nothing you are not forced to believe"

   Default Negation $not\ b_i$ is called **default negation** or **negation as failure** and is often read as "**it is not believed that** b_i **is true**" and this does not imply that b_i is believed to be false.
   Example: p(a) $\leftarrow not\ q(a)$. " If $q(a)$ does not belong to your set of beliefs, then p(a) must".
   In this example, no other rule of the program has $q(a)$ in its head and hence, not thing forces the reasoner, which uses the program as its knowledge base, to believe $q(a)$.

Fig. 3 Semantic web stack.
(From: https://en.wikipedia.
org/wiki/
Semantic_Web_Stack)

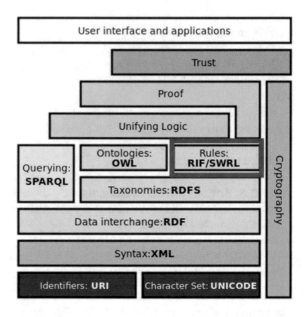

3 Rule Interchange Format (RIF)

The Semantic Web Stack as shown in Fig. 3 is made up of many blocks among which the **rules** block which in the first place extends the ability of web ontology language OWL to describe relations more efficiently [13]. Moreover, these rule languages give means to share and reuse rules between systems. These endowments are important points for expertise processes because they provide it with the capability of interoperability and facilitate collaboration between experts.

Rule Interchange Format (RIF) is a rule language developed by W3C as a recommendation that defines a set of dialect (family of language). In addition to addressing the need for exchange of rules within rule based systems as shown in Fig. 4, RIF in addition facilitates rule-set integration and synthesis [14].[1] It is part of the semantic web infrastructure along with semantic web rule language SWRL 3 which its own patterned scheme on OWL and makes use of it sub-languages (OWL-DL, OWL-Lite). RIF uses XML to provide a syntactic and semantic preserving mapping from one rule base system to another [8]. Indeed, rule-based systems have many differences:

- Paradigms
  Rule languages are built on different approaches: pure first order, logic programming/deductive database, production rules, reactive rules.
- Features and syntaxes
- Commercial interests

---

[1] https://www.w3.org/2005/rules/wiki/Primer.

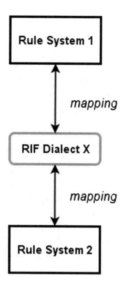

Fig. 4 RIF exchange flow

After the design of the Rule Interchange Format core (RIF-core) dialect, many standard dialects have been created like[2]:

- Basic Logic Dialect (BLD)
  It adds features to the core dialect such as logic function and named arguments
- Production Rules Dialect (PRD)
  It is used to model production rules with features like negation and retraction of facts
- Framework for Logic Dialects (FLD)
  It describes mechanisms for specifying the syntax and semantics of logic RIF dialects, can be used to create some non-standard dialects.

For the case of ASP, one dialect is appropriate even though it is not a standard dialect. This dialect is called Core Answer Set Programming Dialect (CASPD).[3] Syntactically, CASPD corresponds to a language of rules, without function symbols, where rule heads contain disjunctions. Compared to RIF, BLD CASPD does not support equality in rule conclusion or function symbols, but it includes syntax for negation as failure and explicit negation. In addition to common syntax to RIF FLD, CASPD uses Objects frames as in F-logic, and the Internationalized Resource Identifier (IRI) as concepts identifiers and XML schema data type.

---

[2] https://en.wikipedia.org/wiki/Rule_Interchange_Format.

[3] http://ruleml.org/rif/RIF-CASPD.html.

4   Case Study

In this illustration example, we will do a car diagnosis in order to find out among some assumptions, which one is the cause of the car problem. Car failures can be divided into three compartments [12]: start-up state, run-stable state, movement-state. For this work, our use case focuses on the start-up state. This state is described by Howe [15] as follow:

1. The insertion of the car key for the engine starting process.
2. The key is turned and the engine starter will supply a certain among of power to the engine.
3. The engine turnover is provoked by a mixture of gas and fuel in the combustion chamber. This last step completes the engine starting process.

Problems at this stage can be caused by faulty car batteries, spark plugs, car starter, or fuel pump, can also be caused by many other things.

4.1   Problem Modeling with ASP

At this level, we will do the modeling of our use case using Core Answer Set Programming Dialect.

Listing 1 Problem modeling and code

```
car_start(X) :– batteries(X, Month), not fuel_pump_bad(X),Month < 10.

slow_start(X) :– not batteries_bad(X), batteries(X, Month),  Month >
    12.

fuel_pump_bad(X) :– heat(X, Temperature), noise_tank(X), sputter(X),
                            pressure(X, Pressure), Pressure < 30,
    Temperature > 60.

batteries_bad(X) :– slow_start(X), dash_light(X),   batteries_corrosion
    (X),                batteries_indicator(X).
```

This program is defined follow:

1. If it is not believed that the fuel pump is bad and it is believed that the batteries are less than 10 months, then it is believed the car will start.
2. If it is not believed that the car batteries are bad and it is believed that the batteries are more than 12 months, then it is believed the car has a slow start.
3. If it is believed that we have heat in the car at a certain temperature, sputter and noise from tank of the car and fuel pressure is less than 30 PSI then it is believed that the fuel pump is bad.

4. If (it is believed that the car has slow start and the dash_light is on and there is corrosion on batteries and the light of batteries signals when driving) then, it is believed that batteries are bad.

4.2 Formatting ASP to CASPD

Listing 2 Formating use case from ASP to CASPD

```
Document(
Prefix(car <http://www.lgp.enit.fr/concepts#>)

Group (
Forall ?X ?Month(
car:car_start(?X):- And( car:batteries(?X ?Month)  Naf car:
    fuel_pump_bad(?X) ?Month < 10
)

Forall ?X ?Month (
    car:slow_start(?X) :- AndNaf car:batteries_bad(?X) car:batteries(?X
        ?Month)  ?Month > 12)
)

Forall ?X ?Temperature ? Presure (
car:fuel_pump_bad(?X) :- And( car:heat(?X ?Temperature) car:noise_tank
    (?X) car:sputter(?X) car:pressure(?X ?Presure) ?Presure < 30 ?
    Temperature > 60)
)

Forall ?X (
car:batteries_bad(?X) :- And(car:slow_start(?X) car:dash_light(?X) car:
    batteries_corrosion(?X) car:batteries_indicator(?X))
)

car:slow_start(car)
car:dash_light(car)
car:batteries_corrosion(car) car:batteries_indicator(car)
))))
```

5   Conclusion

A guided expertise process is of great importance, but it is also useful to take in consideration exchanges between systems during collaborations among experts. Enhancing the syntax and semantics of the formal representation used for the expertise process can increase the efficient of expertise reasoning. Focusing on these aspects, we engage two formal modeling elements that handle best these important points in the expertise process by using Answer Set Programming and Rule Interchange Format.

We started by presenting an overview of how expertise processes are carry out regarding the NF X50-110 standard, followed by Answer Set Programming and the Rule Interchange Format. ASP declarative modeling 1.1 is an important approach for problem solving because it allows a substantial reduction of implementation and maintenance costs as well as the enhancement of user interactions [16]. It is suitable for expertise since it supports reasoning in situations where new information can lead to contrary decisions or where reasoning is based on incomplete information [17].

A case study using the proposed ASP model is done with a car diagnostic problem available and shareable to other systems, using the W3C recommendation for rule interchange called the Core Answer Set Programming Dialect (CASPD). This rule interchange format, in addition to its support for exchanges among systems, has another important advantage which is its readability from the human point of view. It will be important to extend the Answer Set Programming language to support uncertainty based on belief functions, but also to extend CASPD to consider belief functions.

References

1. Llamas, V. M., Coudert, T., Geneste, L., Bejarano, J. R., & De Valroger, A. (2016). Experience reuse to improve agility in knowledge-driven industrial processes. In: *IEEE International Conference on Industrial Engineering and Engineering Management (Bali)*.
2. Bromme, R., Rambow, R., & Nückles, M. (2001). Expertise and estimating what other people know: The influence of professional experience and type of knowledge. *Journal of experimental psychology: Applied, 7*(4), 317.
3. Grundmann, R. (2017). The problem of expertise in knowledge societies. *Minerva, 55*(1), 25–48.
4. 2013 AFNOR. *Norme NF X50-110 - Qualité en expertise:prescriptions générales de compétences pour une expertise*. AFNOR, 2013.
5. Gebser, M., Kaminski, R., Kaufmann, B., & Schaub, T. (2012). Answer set solving in practice. *Synthesis Lectures on Artificial Intelligence and Machine Learning, 6*(3), 1–238.
6. Brewka, G., Eiter, T., & Truszczyński, M. (2011). Answer set programming at a glance. *Communications of the ACM, 54*(12), 92–103.
7. Law, M., Russo, A., & Broda, K. (2018). The complexity and generality of learning answer set programs. *Artificial Intelligence, 259*, 110–146.
8. Shen, Y.-D. & Eiter, T. (2019). Determining inference semantics for disjunctive logic programs
9. Janssen, J., & Schockaert, S., Vermeir, D., & De Cock, M. (2012) *Answer set programming for continuous domains: A fuzzy logic approach* (Vol. 5). Springer Science & Business Media.

10. Kakas, A. C. (1994). Default reasoning via negation as failure. In *Foundations of Knowledge Representation and Reasoning* (pp. 160–178). Springer.
11. Gelfond, M. & Kahl, Y. (2014) *Knowledge representation, reasoning, and the design of intelligent agents: The answer-set programming approach*. Cambridge University Press.
12. Al-Taani, A. T. (2005). An expert system for car failure diagnosis. *IEC (Prague), 5*, 457–560.
13. Rattanasawad, T., Saikaew, K. R., Buranarach, M., & Supnithi, T. (2013). A review and comparison of rule languages and rule-based inference engines for the semantic web. In *2013 International Computer Science and Engineering Conference (ICSEC)*, pp. 1–6. IEEE.
14. Kifer, M. & Boley, H. (2013). Rif overview. W3C working draft, W3C (October 2009). http://www.w3.org/TR/rif-overview
15. Howe, J. (2019). How to troubleshoot a car that is hard to Start. *yourmechanic.com/article/how-to-troubleshoot-a-car-that-is-hard-to-start-by-jessica-howe*.
16. Falkner, A., Friedrich, G., Schekotihin, K., Taupe, R., & Teppan, E. C. (2018). Industrial applications of answer set programming. *KI-Künstliche Intelligenz, 32*(2–3), 165–176.
17. Nicolas, P., Garcia, L., Stéphan, I., & Lefèvre, C. (2006). Possibilistic uncertainty handling for answer set programming. *Annals of Mathematics and Artificial Intelligence, 47* (1–2), 139–181.

Metadata for Complementing Standards and Formalisation of the Technical Reserve Calculation

Giusty Guerrero, Ana X. Halabi-Echeverry ⓘ, and Juan C. Aldana ⓘ

Abstract The Colombian health sector is comprised among other entities by Health Promoting Entities (EPS) which currently face several challenges given the complexity of the activities they carry out with their insurance role. Innumerable problems with the service promise and poor-quality information used for analysis, management and decision-making treat their sustainability. The information processes of the EPS are focused on collecting the information for the calculation of the Technical Reserve (provisions) according to a regulated period. The formalisation of the information the EPS produced from several repositories is often mislead or deceived. This paper presents a rudimentary metadata for complementing standards and formalisation of the Technical Reserve calculation, underlying one of the most recurrent standards for this aim 'the HL7 FHIR standard (release 4)'. Conclusions point to an evidence that the Colombian health system is in the early stages of data management. It is important to begin by associating standards such as the HL7-FHIR with the local requirements allowing harmonised and seamless communication between the parts of the system. Similarly, opportunities exist to make available the HL7-FHIR standard to facilitate the calculation of the Technical Reserve in Colombia and other countries of the region.

Keywords ETL formalisation · HL7 FHIR standard · Public Health Organisations · Policy and regulation

G. Guerrero
Universidad de La Sabana, Autop. Norte de Bogota km 7, 1400013 Chía, Colombia

A. X. Halabi-Echeverry (✉)
Research and Advise, NextPort vCoE INC., 2113 North Ryde, Australia
e-mail: axhalabi@nextport.com.co

J. C. Aldana
Universidad Nacional de Colombia, Bogotá, Colombia
e-mail: jcaldanab@unal.edu.co

© The Author(s), under exclusive license to Springer Nature Switzerland AG 2023
B. Archimède et al. (eds.), *Enterprise Interoperability IX*, Proceedings of the I-ESA Conferences 10, https://doi.org/10.1007/978-3-030-90387-9_21

1 Introduction

In public entities, many of the inputs/data needed are either missing, incomplete, inconsistent or not taken into consideration. If there is a lack of information or data, this data may use different labels/names, units of measurement and time frames. Among the main obstacles presented are frequent problems of duplication of documentation mainly related to not having clear information controls between the different processes that produce the data. The latter causes loss of users, increase in demands for the non-provision of services, economic fines to fail with the requirements. Particularly, for Health Promoting Entities (EPS), Resolutions 4175 de 2014 and 412 of 2015 issued by the National Superintendence of Health—Supersalud in Colombia [1, 2] indicate that "Public entities authorised to operate health insurances may use technically recognised methods or procedures to calculate provisions. For this purpose, the data must be used with the authorisation of the National Superintendence of Health" [2], p. 3. Decree 2702 of 2014, underlines obligations and established guidelines to account each type of provision, which are described in Article 3 of Resolution 412 of 2015. All of which set up the culture and good practices to control the development of the information, cycle management and validation capacity as urgent matters [3].

One of the most frequent problems is the absence of a structured and integrated model for the information produced by the different entities. Thalheim [4, 5] addresses the need for data formalisation, avoiding its use just as an arbitrary source of information. Bustamante et al. [6] state that ETL (extraction, transformation and loading of information) is one of the critical activities involved in having consistent, uniform and available data. ETL processes are modelled by visual representations. The ETL phases according to Casters et al., [7], p. 5 are: (a) Extract: All processing required to connect to various data sources, extract the data from these data sources and make the data available to the subsequent processing steps, (b) Transform: Any function applied to the extracted data between the extraction from sources and loading into targets and (c) Load: All processing required to load the data in a target system. This part of the process consists of a lot more than just bulk loading transformed data into a target table. Duque et al. [8] develop an ETL model to extract data from various sources through filtering and error detection, with the objective of guaranteeing the integrity, consistency and quality of the stored data. The model has several phases to be executed in web environments. For the detection, the generated errors are identified as outliers and inconsistent data; this is possible thanks to the use of predefined comparison tables better known as "tbl-variables". The detection of errors and inconsistencies allows the correction of the data according to specific standards. The final phase allows to migrate the data to new tables known as "tbl-temporal" in a definitive repository. Muñoz et al. [3] mention the development of a model for the integration and consultation of information by referring to the description of the information exchange process, which is possible due to the creation of information exchange files. The process is supported through ETL in Pentaho Data Integration Community Edition 7.1. The results notice the integrated and consolidated information in a single repository, helping decision-makers access it in real time.

The data standardisation solves the problem of the information required for the calculation of the Technical Reserve and includes the possibility of taking the information from different data sources. The ETL formalisation is one of the critical activities involved in having consistent, uniform, and available data standardisation to inform policy, and regulation; this guarantees the reliability, integrity, and consistency of data. A flow of information as a process for accounting the Technical Reserve from the starting point of signing contracts to reimbursements for services generated to users, need elements that constitute metadata for Public Health Organizations.

On the other hand, HL7 FHIR is a standard for healthcare data exchange. It is based on ISO/HL7 21731:2014 Health informatics—Release 4. This standard is a well-known documented standard with several modules among which is the financial module supporting the resources and services of the entities, such as costing, financial transactions, and billing within a healthcare provider and an insurer or patient [9, 10]. In particular, the HL7 FHIR standard promises healthcare interoperability through more detail conceptualisations which can be stored as ontologies that help to overcome issues of misleading meanings and interpretation [9, 10]. However, the HL7 FHIR standard does not provide the actuarial calculation for the Technical Reserve, in part due to the strategic approach holded by health entities.

Other standards are: (a) ISO 13940 to support continuity of care, (b) ISO 18308:2011 for an electronic health architecture, (c) ISO 22600 for maintaining, management and access control of the system, (d) ISO/IEC 10746-1:1998 for an open distributed processing, (e) ISO 27789:2013, for audit trails of records, (f) ISO/IEC/IEEE 24765:2017 and ISO/IEC 2382:2015 for vocabulary use and (g) ISO 8601:2004 for data elements and interchange formats.

To sum up, none of the previous standards are ample and sufficient to frame the Technical Reserve as a complex and actuarial calculation local policy and regulation.

The research question stated in this paper is as follows:

What can be the metadata for complementing standards and formalisation of the Technical Reserve Calculation?

Section 2 presents an application case in Colombia to allow for a possible solution to the information harmonisation problem on public data of health organisations to inform policy and regulation.

2 Application Case: Public Data of Health Organisations in Colombia

The Colombian health sector is comprised among other entities by Health Promoting Entities—EPS—which currently face a series of strategic, tactical, and operational challenges given the complexity of their activities, the insurance role and daily shortcomings to manage the health of users. Therefore, to maintain an adequate planning, control, and execution of their operations, budgeted resources must be executed in less time; being an example of efficient institutions. One of the critical issues that currently affect EPS is the capture and standardisation of the information produced; which generates problems with the service promise and poor-quality information (the final value chain product of a health promoting entity) used for analysis, management, and decision-making. According to a report in [11] issued by the National Superintendence of Health, there is evidence of the total sanctions imposed on institutions with non-compliance records and avoidance in obliged information. The report reached an estimated value of two billion Colombian pesos distributed in 45 health entities [12], p. 31.

Currently, to comply with Decree 2702 of 2014 and Resolution 412 of 2015, the information of the Health organisations are focused on collecting the information for the calculation of the Technical Reserve (provisions) according to a regulated period. Several issues arise due to the poor interaction between the processes that produce the information for the calculation of the Technical Reserve, such as the low percentage of compliance with the parameters or characteristics of data, the gap between the dates in which the delivery of reports must be made [13], and the lack of the agreements for controlled services between the different processes [14, 15]. Under Resolution 412 of 2015, the Technical Reserve is the measure that "allows determining the capacity of the health entities to operate the health insurance to make in front current or eventual obligations contracted by their activity and constitute as the main source to attend the payment of the health services" [2], p. 2. This points to the fact that the Technical Reserve are resources that an EPS must ensure in order to be able to respond to all the commitments acquired allowing the continuity of their operations. Cuevas [16] states that the Technical Reserve refers to the resources that an entity allocates to support its obligations with its suppliers, so it is necessary to identify and quantify what are the obligations derived from the contracts signed for accounting purposes. On the other hand, the norm indicates the calculation required by the Technical Reserve in Decree 2702 of 2014, and those must be known and followed by the entity.

Section 3 presents the identification of the information problems in the case study at hand.

3   Publicly Available Information

In accordance with the provisions of the entity, the National Superintendence of Health requests the information that should support the calculation of the *Technical Reserve* [12], p. 1; and must be submitted by the Health Promoting Entities—EPS, annually by means of electronic formats.

The general structure of the public information for the calculation of the *Technical Reserve* is set periodically in accordance with the provisions of Decree 2702 of 2014 and the Resolution 412 of 2015 [17, 18]. The information generated by the EPS can be traced in nine databases, [12], p. 1. The information within the contracts and authorisation databases issued by the EPS are the basic information to validate the registered fields of capitas and/or packages not invoiced. The information within the payment details (total or partial) validate the registered fields of capitas and/or packages paid. In case of inconsistencies, the information indicates if the provider is not hired or the service provided not authorised. The detailed information follows:

(a) Instances of the EPS contracts also signed by the Health Delivery Institutions—IPS—and pharmacies
(b) Instances of the authorizations executed by the EPS with its associates
(c) Instances of invoices charged by the IPS to the EPS for generated services
(d) Instances of authorized disabilities for associates who require reimbursements.

Table 1 shows available information for the *Technical Reserve* from the starting point of signing contracts to reimbursements for generated services of users.

3.1 Types of Inconsistencies

Public entities use inputs/data either missing, incomplete, inconsistent or not taken into consideration. If there is a lack of information or data, this data may use different labels/names, units of measurement and time frames. Among the main obstacles are frequent problems of duplication of documentation not related or having clear information between the different processes that produce the data. This causes loss of users, increase in demands for non-provision of services and economic fines to comply with the requirements of the regulations. Among the different inconsistencies are:

- **Number of characters**: this error occurs when the number of characters is not consistent or does not meet the total characters defined by the source of the data, for instance, a tax identification number must contain exactly nine digits otherwise an error occurs.
- **Special characters**: This error happens when different characters are added to others predetermined, for instance, the use of points and commas with the ID number: 66.851.127
- **Complementary Characters:** the error occurs when specified complements were not assigned to initial characters. For instance, the passport number must be accompanied by numbers and letters such as CC66851127. In case of not assigning the complement, the character would not be easily classified.
- **Acronyms or identification characters**: this error occurs when the meaning of an acronym has not been identified, is not accepted or is misused.
- **Date formats**: This error occurs when the date format is not met, this cause an error in the analysis when comparing with other formats.
- **Hierarchical or logical order in data**: This error occurs when the information is not organised by date creation or date generation, also by alphabetical order,

Table 1 Available information for the *Technical Reserve*

Database	Description	Agent
Contract details	Corresponds to the written agreements established between the EPS and the different companies that provide health services (IPS) and/or people in the sector, for the provision of health services, to EPS associates	EPS
Authorization details	Corresponds to the authorizations issued by the EPS, for the provision of health services to its associates in its service delivery network, during the 12 months prior to the cut-off date and from which no invoice has been received	EPS
Details of capitas and/or packages not billed	Defines the value of the monthly contracts established with the current health service providers, for which no invoice has been received and therefore have not been paid	EPS
Detailed of invoices informed	Corresponds to the invoices' value for the provision of health services in the network of the EPS providers, that have not been paid in full at the cut-off date. For its calculation, all invoices pending of payment (partial or total) must be included, regardless of the date of reporting	EPS
Detail of capitas and/or billed packages	Defines the value of the monthly contracts with current health service providers, for which invoices were received and have been paid in full at the cut-off date	EPS
Payment detail (total or partial)	Corresponds to total or partial payments of the invoices received from health service providers, made up to 48 months before the cut-off date	EPS
Details of capitas and/or packages paid (total or partial)	Defines the value of the monthly contracts established with the current health service providers, for whom invoices were received and have been paid in whole or in part, up to 36 months before the cut-off date	EPS
Details of disabilities paid (totally or partially)	Corresponds to the value of the disabilities that users have requested, through health service providers, which have been paid in full or in part for up to 48 months prior to the cut	EPS
Details of recognized disabilities	It corresponds to the value of the disabilities that the users have processed at the EPS, and that at the cut-off date have not been paid in full	EPS

ascending or descending order or according to the predefined guidelines and level of importance.

- **Negative or positive data, blank or zero**: this error occurs when negative or positive values are assigned to data but are not consistent with the type of data, or when no information is assigned to the field or found in zero.
- **Currency symbols and abbreviations**: This error occurs when in some values the appropriate currency or abbreviation configuration is not met, which generates the significant alteration in the values.

4 Establishing Rudimentary Metadata for Complementing Standards Development on Technical Reserve Calculations

The ELT data formalisation solves the problem of the information required for the calculation of the Technical Reserve and includes the possibility of taking the information from different data sources, which is vital, because they are located in different databases. The ELT data standardisation enables policy, regulation and business rules validation which allow for the detection of errors and the correction of them when possible, this guarantees the reliability, integrity and consistency of data.

An ELT data process is presented in Fig. 1. It is important to specify that the implementation of an ETL process reduces the risk of information processing by several agents, due to its automatisation, traceability and security of data. One of the stages in Fig. 1 mentions the elements that constitute metadata for Public Health Organisations. Subsection 4.1 establishes rudimentary metadata, not meant to be safe for implementation but that give insights for complementing standards development on the *Technical Reserve* calculation.

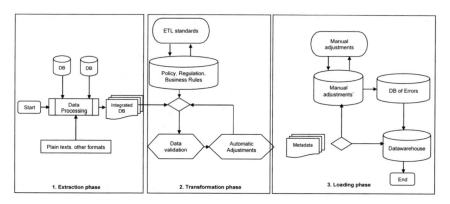

Fig. 1 ETL formalisation for an EPS

4.1 Metadata Expressions

Tax identification number (TAX.ID.AGENT) is a 9-digit code valid in Colombia for a business or individual. There are legacy aspects attached to it. There must not be obsolete tax codes cases, otherwise, they need to be removed. Unidentified in HL7-FHIR.

Special record for IPS qualified by MSPS (REPS.ID.IPS) is a number or 12-digit code that corresponds to an special database record for Health Service Providers. There are legacy aspects attached to it. Special cases between 9 and 12 digits are possible. If a health entity has not the right to operate authorised by the Ministry of Health and Social Protection, its REPS can be 12-digit code of zeros. Unidentified in HL7-FHIR.

Tariff manual (TARIFF. MANUAL) is a unique character representing obsolete or current tariff guidelines. There are legacy aspects attached to it and indicating the date for the tariff validation, such as ISS (Social Security Institute years 2000, 2001 and 2004, SOAT (Mandatory Traffic Accident Insurance) years 2014 to 2017, and other guidelines such as the baseline tariff guideline. Unidentified in HL7-FHIR.

Agreement validation of terms (AGREEM. VALIDATION) is a unique character representing the state of agreements: value N for current agreement terms, R for previous agreement terms and O for agreement adjustments or modifications. Unidentified in HL7-FHIR.

Date format (DATE. FORMAT) admitted date entry stored as a text. Is a 10 string/number in a format DD/MM/YYYY.

CUPS (CUPS. SERVICES) is a unique 6-digit code to denote health service classifications. It is an ID assign to service request HL7-FHIR.

CUMS (CUMS. MEDICATION) is a unique 4 to 9-digit code to denote medication classifications. It is an ID assign to medical products HL7-FHIR.

Base tariff (BASE.TARIFF) is a positive number corresponding to the tariff of the providers. It is given in Colombian pesos ($COL) and must be greater than zero. Part of HL7-FHIR defines the minimum tariff of services to be provided.

Rate surcharge (TARIFF. RATE SURCHARGE) is a percentage which denote an increment over the tariff guidelines. Unidentified in HL7-FHIR.

Updated tariff (UPDATE.TARIFF) is the value corresponding to the comparison of baseline tariffs and the surcharges. It is obtained from the formula: baseline tariff * (1 + % rate surcharge), given in Colombian pesos ($COL), and must be greater than zero. Unidentified in HL7-FHIR.

Authorization number (AUTHORIZATION.NUMBER) is an authorization identifier of 8-digit code endorsed by the EPS and given by the providers to users accessing the service (HL7-FHIR).

Scope authorization (SCOPE.AUTHORIZATION) is a unique character of the type of service provided: A for Outpatient, H for Hospital, U for Emergency and D for Home. It is taken as a preauthorization code (HL7-FHIR).

Unit of measure (UNIT.OF.MEASURE) is a 60-string character denoting the medication unit. It is the minimum amount ordered by the physicians or distributed by pharmacies such as a bottle, a box and a number of drops. It also related to the medicine product packaged (HL7-FHIR).

Authorized quantity (AUTHORIZED.QUANTITY) is a positive number greater than zero that indicates authorized services and medicines. It obtained from doctors' referrals (HL7-FHIR).

Tariff estimation (TARIFF. ESTIMATION) is a unique character representing tariff standards between regulated tariff guidelines and tariffs agreed between EPS and IPS. It can be: T were the tariff is negotiable and P were the tariff is the average value in the market. Unidentified in HL7-FHIR.

Medication value (MEDICATION.VALUE) is a monetary unit greater than zero indicating the amount authorised by the EPS. It is given in Colombian pesos ($COL). Unidentified in HL7-FHIR but it if in relation to the cost of a medication then is found in HL7-FHIR.

Contract value month (CONTRACT.VALUE.MONTH) is a monetary value greater than zero that corresponds to the services agreed by the EPS and IPS for a given month. It is a part of the contract among the parties (HL7-FHIR).

Agreement record (TAX_AGREM. TYPE_IPS_EPS) is a single string code that represents the TAX 9-digit code followed by a hyphen and 3 to 6 alphanumeric characters. The alphanumeric characters denote the type of concluded agreement, i.e., EVENT: high amount of health services, CAPITA: regular health services, PAF/PG: special health services. For instance: 800000545_CAPITA (HL7-FHIR).

Days of disability (DAYS.DISABILITY) is the number of days given to a patient for diagnosis and treatment. Unidentified in HL7-FHIR, but if related to results of diagnostic report is found in HL7-FHIR.

Basic affiliation income (BASIC.AFFL.INCOME) basic salary that affiliates report to the health system. Unidentified in HL7-FHIR.

Glossy value (GLOSSED.VALUE) is an stipulated loss value (zero is possible), which corresponds to invoices in an approval process but with some inconsistencies. If it is a part of a payment notice is found in HL7-FHIR.

Glossy status (GLOSSED.STATUS) 1-digit code with two possible outcomes: 1 for reconciled invoices and 2 otherwise. It is an ID for payment reconciliation (HL7-FHIR).

Technical reserve medication unitary value (TEC.RESERVE.MEDICATION.

UNIT.VALUE) is the value corresponding to the calculation between the medication unit and the average monetary value for this unit. That also be denoted as unitary technical reserve authorized by the EPS. It is given in Colombian pesos ($COL). If it is a part of the medication dispense is found in HL7-FHIR.

Value of the Known Technical Reserve (VALUE.KNOWN.TEC.RESERVE) is the monetary value corresponding to the calculation of the paid invoices and glossy invoices, corresponding to services, medications and disabilities. It is given in Colombian pesos ($COL) and must be greater than zero. If it is a part of payment reconciliation is found in HL7-FHIR.

Value of the Unknown Technical Reserve (VALUE.UNKNOWN.TEC.RESERVE) is the monetary value corresponding to the calculation of the unpaid invoices and glossy invoices because they have not been reconciliated. It is given in Colombian pesos ($COL) and must be greater than zero. If it is in relation to payment notice is found in HL7-FHIR.

Other reserves (OTHER.RESERVES) are the provisions that the EPS must make for losses in its operation. Unidentified in HL7-FHIR.

Technical Reserve (TECHNICAL.RESERVE) corresponds to the capacity of the EPS to guarantee the health insurance operation against current and incurred obligations. They constitute the well-known: Value of the Known Technical Reserve, Value of the Unknown Technical Reserve and Other reserves. Unidentified in HL7-FHIR.

5   Informed Policy and Regulation

Informed policy and regulation are dependent on the data at hand. Of relevance is that local data extraction can be used for some instances but rarely used in general contexts of decision-making, i.e., an entity of control may not see the data problem such as is the data duplication problem. Another example is given when recording an agreement. Several records may appear between more than one EPS and IPS, complicating the decision-making process. Although ELT data standardisation allows to overcome certainly the problem, it also anticipates pressures among the parts in pursuing the same business rules and negotiations. An ELT data standardisation is more than a technological process, it is capable of accomplishing policy and regulation, from which it takes the informed rules of the system.

Moreover, the ETL formalisation of the Technical Reserve considers the information omitted by the health institutions (belonging to the Mandatory Health Plan—POS-, or the complementary disability health system). In accordance with the provisions of Resolution 412 of 2015 issued by the National Superintendence of Health—Supersalud, the Technical Reserve is the unique measure to ensure that an EPS adequately manages its economic resources to respond to the acquired commitments. That may perpetuate the provisions and ease the management of the population health. According to Cuevas [16], identifying and quantifying those obligations derived in transparent contracts and an accounting records guaranteeing the payment and future recognition of the national health system (as it is indicated in Decree 2702 of 2014).

The metadata presented gives an indication that the produced and shared of information in the Colombian health system is still in early stages. While there is no a unified data framework for the health management system, it is relevant to associate standards such as the HL7-FHIR to local requirements allowing harmonised and seamless communication between the parts. Similarly, opportunities exist to make available the HL7-FHIR standard, as an online, easy-to-use tool that may facilitate the calculation of the Technical Reserve in Colombia and other countries of the region.

6 Conclusions

As shown, ETL is a mandatory standardisation process for organisations of all kind. Data consolidation, cleaning, transformation and loading into a continuous and rigorous process allows decision-making free of errors, creating opportunities for management and governance. High data volume can be of interest for the business sustainability, but it requires the knowledge to transform raw data into smart pieces of information.

The public health sector in Colombia is more than any other in need of ETL formalisation due to its complicated data management and its use of public sources which impact a significant number of users in the system. Similarly, opportunities exist to make available the HL7-FHIR standard to support health entities not only from the transactional operation side but also to the sustainable approach of the system.

The importance of the information exchange between EPS and IPS lies in the surveillance and control of these entities providing alerts to the national government and anticipating sustainable decisions. This would guarantee constitutional compliance indicating that health must be a right and must comply with the protection of its entire population. In Colombia, the Technical Reserve is the unique measure to ensure that an EPS adequately manages its economic resources to respond to the acquired commitments, perpetuating the provisions and management of the health for its patients.

References

1. SNS - Superintendencia Nacional de Salud (2014). Resolucion 4175 de 2014, Retrieved from https://normativa.colpensiones.gov.co/colpens/docs/resolucion_supersalud_4175_2014.htm
2. SNS—Superintendencia Nacional de Salud (2015). Resolucion 412 de 2015, 13. Retrieved from https://docs.supersalud.gov.co/PortalWeb/Juridica/Resoluciones/R_2015_Norma_000 412.pdf

3. Muñoz, V. D., Garcia, H., Rubiera, O., López, C. R., & Wilford, I. (2015). Nuevas capacidades de integración de información docente en Instituciones de Educación Superior en Cuba. *Ciencias de La Información, 46*(2), 3–8.
4. Thalheim, B. (2010). Data management: Challenges and solutions. *A critical reflection of data management practices in research projects* (pp. 40).
5. Thalheim, B. (Ed.). (2007). *Achievements and problems of conceptual modelling*. Springer-Verlag.
6. Bustamante, A., Galvis, E. A., & Gómez, L. C. (2013). Técnicas de modelado de procesos de ETL: Una revisión de alternativas y su aplicación en un proyecto de desarrollo de una solución de BI. *Scientia et Technica, 18*(1), 185–192.
7. Casters, M., Bouman, R., & Dongen, J. (2010). *Pentaho kettle solutions: Building open source ETL solutions with pentaho data integration*. Wiley.
8. Duque, N. D., Hernández, E. J., Pérez, Á. M., Arroyave, A. F., & Espinoza, D. A. (2016). Modelo para el Proceso de Extracción, Transformación y Carga en Bodegas de Datos. Una Aplicación con Datos Ambientales. *Ciencia e Ingeniería Neogranadina*, 26(2), 95–109.
9. HL7 FHIR Release 4, January, 20 (2020). *Re: Financial module*. Retrieve from http://hl7.org/fhir/StructureDefinition/cqif-citation
10. Kiourtis, A., Nifakos, S., Mavrogiorgou, A., & Kyriazis, D. (2019). Aggregating the syntactic and semantic similarity of healthcare data towards their transformation to HL7 FHIR through ontology matching. *International Journal of Medical Informatics*, 132, 104002. https://doi.org/10.1016/j.ijmedinf.2019.104002
11. SNS—Superintendencia Nacional de Salud (2017). Informe De Gestión 2017. Supersalud, 2017(COFL02), (pp. 1–78). Retrieved from http://www.prosperidadsocial.gov.co/ent/gen/trs/Documents/Informe de Gestión ANSPE 2015- Cierre contable.pdf
12. SNS—Superintendencia Nacional de Salud (2018). Detalle de los insumos, listados o bases de datos que soportan el cálculo de la reserva técnica. Remitido por Supersalud en código radicado 2-2018-042407.
13. Dueñas, M. X. (2009). Minería de datos espaciales en búsqueda de la verdadera información. *Universidad Francisco José de Caldas, 13*(1), 137–156.
14. Calabria, J. C. (2011). Construcción y poblamiento de un datawarehouse basado en el paradigma de bases de datos objeto relacional. *Revista Prospect, 9*(1), 69–77.
15. INVIMA. (2018). Listado Único de Medicamentos Vigentes a agosto 2018. Retrieved from https://www.invima.gov.co/rss/213-tramites-y-servicios/consultas-registros-y-documentos-asociados/806-listado-codigo-unico-de-medicamentos.html
16. Cuevas, M. C. (2011). El Régimen de Reservas Técnicas en Colombia, Retos Futuros. Fasecolda, 494–575
17. MSPS—Ministerio de Salud y Protección Social. Decreto 2702 de 2014, 13 (2014). Retrieved from http://wp.presidencia.gov.co/sitios/normativa/decretos/2014/Decretos2014/DECRETO 2702 DEL 23 DE DICIEMBRE DE 2014.pdf
18. MSPS—Ministerio de Salud y Protección Social (2018). Registro Especial de Prestadores de Servicios de Salud. Retrieved from https://www.datos.gov.co/Salud-y-Protección-Social/Registro-Especial-de-Prestadores-de-Servicios-de-S/c36g-9fc2/data

Business Oriented Applications

A Framework to Formulate Models and Identify Algorithms to Solve Large-Sized Industrial Planning Problems

Eduardo Guzman, Beatriz Andres, and Raul Poler

Abstract This work puts forward a framework that guides model designers about the formulation and notation of mathematical models to solve replenishment, production and delivery plans. Having characterised plans, this framework generates the set of decision variables, input data and objectives to formulate the defined planning problem. It also identifies the most proper algorithms to solve the previously formulated planning models. The application of algorithms helps to solve large-scale industrial planning problems with a limited computation capacity and extends capabilities beyond solving mathematical programming models.

Keywords Mathematical programming · Model formulation · Solver algorithm · Large-sized planning problems

1 Introduction

The field of study of quantitative methods offers a solution to those problems that emerge in industrial organisation and supply chain management by designing efficient mathematical models and algorithms to deal with decision-making procedures.

Research into planning areas has exponentially evolved since the 1950s, which was when operations research gave the first results promoted by computational complexity improvement and algorithms development to solve large-sized problems [1].

In this context, mathematical models described the problem and provided a closed series of solutions to obtain an optimal or approximate solution by means of which an

E. Guzman · B. Andres · R. Poler (✉)
Research Centre on Production Management and Engineering (CIGIP), Universitat Politècnica de València (UPV), C/Alarcón, 03801 Alcoy, Spain
e-mail: rpoler@cigip.upv.es

E. Guzman
e-mail: eguzman@cigip.upv.es

B. Andres
e-mail: bandres@cigip.upv.es

approximation that moved closer to the true solution was accomplished. However, mathematical models' nature makes modelling realistic highly complex systems difficult [2]. The continuous improvement of computational mathematical programming capabilities has facilitated their solution. Nevertheless, the running times and computational costs to obtain the solutions of very large problems involving many thousands of decision variables and restrictive constraints are still inefficient, and only limited-sized models have been solved to date. In this context, defining and applying heuristics and metaheuristics have improved solving large-scale planning problems with limited computation capacity and extended capabilities beyond merely solving mathematical programming models. Nowadays, the operations management research area offers the proposal of matheuristic as an interoperation of metaheuristics and mathematical programming techniques [3]. The innovative traits of designing matheuristic require of modellers more expertise. Model designers also have to deal with highly complex modelling and must solve planning problems in supply chains, which are characterised by a vast amount of input data and variables, and also by conflicting constraints and objectives appearing among supply chain partners [4]. All in all, through their solution by algorithms, mathematical models help decision-making by generating optimal, or near-optimal, solutions according to an established objective.

In order to confer the design of mathematical models and algorithms a higher level of familiarity, this paper proposes a framework to guide: (i) the formulation and notation of the models used to solve supply chain planning problems, including source, make and delivery, by employing the plans defined by [5, 6]; (ii) the identification of algorithms so they are more properly used to solve the previously formulated planning model. In any case, if users are interested in building models and algorithms, they have to define the type of plan to be solved and the horizon. Having characterised the plan, the framework herein proposed allows the generation of a set of decision variables, input data and objectives to formulate the defined planning problem.

This paper is arranged as follows: Sect. 2 offers a literature review of the mathematical modelling approach and works formerly proposed to facilitate model designers' task of formulating mathematical models to support decision-making in the planning context. Section 3 contains the main contribution: a conceptual framework to formulate planning models and to identify solver algorithms. Section 4 presents the case study by applying the proposed conceptual framework to formulate a mathematical model by employing a real planning problem from a second-tier automotive supplier. Finally, Sect. 5 includes the conclusions.

2 Literature Review

According to Christou [1], planning processes are a focal point of enterprises and SC operations, and one of the most significant activities in industrial organisation. The operations management research area provides techniques and methods to model planning processes. This makes operations research a discipline that can deal with the application of advanced analytical and mathematical methods, theories and techniques to support the decision-making process in supply chains. Some examples of such are business analytics, computer science, decision analysis, forecasting, game theory, graph theory, industrial engineering, logistics, mathematical modelling, mathematical optimisation, probability and statistics, simulation, stochastic processes and supply chain management.

In the supply chain and industrial management area, operations research deals with determining the extreme values of planning processes objectives, e.g. maximisation or minimisation. When a researcher or an industrial expert formulates planning models, (s)he cannot always be able to exactly depict the organisation's reality. Instead the person in charge of modelling, the problem should have to simplify it to make it solvable after selecting the solver algorithm.

According to Pidd [7], "*a model is an explicit and external representation of part of the reality as it is seen by people who want to use the model in order to understand, change, manage and control that part of reality*". Models represent part of reality. However, reality is always more complex than any model, regardless of how sophisticated it might be. The model designer has to determine which aspects are relevant, and which are not, depending on the objective intended to be fulfilled. Experience shows that the main benefit from generating a model is to understand what the modeller acquires from reality's behaviour. Quite often when developing a model, the designer becomes aware of information that (s)he has never paid attention to. Moreover, it is quite usual that, when a modeller formulates the model, real and contradictory data appear between different elements of reality. In his book "Quantitative Methods in Supply Chain Management: Models and Algorithms", Christou[1] provides an example of what would occur when modelling a job-shop scheduling planning problem: "*... almost all of the hard constraints we shall encounter in job-shop scheduling and due-date management, in reality are not that "hard" but are soft constraints in that often, violating one of them by a small slack does not violate any physical laws nor does it hurt company profitability in the long run*".

Bearing all this in mind, the reviewed literature clearly shows the complexity of formulating a model from scratch. Some authors have proposed methodologies and tools that efficiently deal with modelling planning processes or have provided a realistic formulation with knowledge-based tools that help non-expert users to build mathematical models in different planning areas. For this purpose, different papers in the literature have been identified. Hackman and Leachman[8] introduce a general framework that guides the management scientist's formulation of deterministic models of production processes. The work of Krishnan [9] proposes a knowledge-based tool for building the algebraic schema of appropriate linear programming (LP) models for production, distribution and inventory (PDI) planning problems. Krishnan [10] studies the application of knowledge-based techniques to support various modelling process phases by integrating artificial intelligence (AI) techniques into decision support systems (DSS).

Shapiro[11] classifies models according to the effect their result has at the normative or descriptive level. Mathematical models are normative (in turn they can be classified as optimisation models and resolution models by heuristics). Descriptive models cover all the modelling techniques that do not involve defining mathematical structures that, in turn, define a desirable solution to be implemented. Before going further into the use and formulation of mathematical models, it is worth clarifying that the literature addresses the task of modelling planning problems from a perspective that is not only normative, but is also descriptive and conceptual. Indeed Hernández et al., [12] state that the conceptual model is helpful for gaining a better understanding of the system and, consequently, of detecting irregularities and suggesting improvements. Accordingly, Hernandez et al. [13] propose a conceptual model for the production and transport planning process in the automobile sector.

Although the authors of the present paper are aware of the relevance of other modelling approaches, the work herein conducted focuses on planning process modelling from a normative perspective. In an attempt to facilitate the representation of planning problems, Hashimoto and Kubo [14] collect a set of fundamental mathematical optimisation models (mixed integer linear programming, MILP), such as logistics network design, inventory, scheduling, lot-sizing, and vehicle routing models, to provide modellers with knowledge about basic mathematical formulations in the enterprise planning context. Therefore, the work of Hashimoto and Kubo [14] gives modellers a clue about the indices, input data, objectives, variables and output data that are widely used to formulate planning problems.

According to Mula et al., [15] the most widespread approach to model planning problems is MILP. Yet some characteristics are identified as limitations when solving planning problems through MILP, especially when considering enterprises' real resolution environments. The main weaknesses are related to: (i) the combinatorial nature of real-world problems, in which the amount of decision variables exponentially increases when the number of plants, products or time periods increases; (ii) the large volume of data. Both cases generate an extensive use of computer memory, which results in an increased need for solution time [16]. It is here when solver algorithms and heuristics come into play to employ them as complementary techniques to solve mathematical programming models, mainly integer linear programming (LP). In line with this, Prasad et al.[16] propose a collection of algorithms to support solving mathematical models, formulated to support planning decision-making.

The literature review clearly indicates that a lot of progress has been made in proposing algorithms to support solving mathematically modelled planning problems (MILP). Nevertheless, the papers analysed in this section focus only on proposing approaches that support LP model formulation, and do not consider the solver mechanisms that solve them. To the best of our knowledge, the works that develop mechanisms to help to formulate mathematical models do not address the identification of appropriate solver algorithms.

In light of this, the present paper proposes a theoretical framework to support: (i) the formulation of mathematical models in the replenishment, production and delivery planning contexts; (ii) the identification of best fitting algorithms that solve real-world planning problems, regardless of the vast amount of data required for the problem to be solved in a real enterprise or supply chain. These algorithms enable an efficient computationally solution process during which a vast amount of data is used.

3 Conceptual Framework to Formulate Planning Models and to Identify Solver Algorithms

When addressing planning processes, it is necessary to develop optimisation and decision support tools that help to explore and analyse alternatives that can optimise economic performance and service levels [17]. The quantitative methods area studies ways to improve the quality, understanding and consequences of the decision-making process. Mathematical programming plays a very important role in this research area.

Planning models imply high complexity levels, especially if models are applied to an enterprise's full-sized planning or a supply chain network. The combinatorial nature of real-world problems makes models exponentially complex in terms of input data, objectives, constraints and decision variables. Consequently, modellers must possess sufficient knowledge and background about the plans to be represented and solved. They must also have enough expertise to mathematically formulate the planning process by considering the soft and hard constraints, as well as a set of input data, required to meet the proposed objective.

So despite making efforts to simplify the planning problem, computationally solving mathematical models is still complex and time-consuming. Although significant progress has been made in the general solving mathematical programming area, current optimisation algorithms are still unsatisfactory for efficiently solving all general medium-sized integer linear programmes in reasonable times. Both complexity and inefficiency increase when solving full-sized planning problems, and this involves large datasets. Although adequate computational techniques have been developed for special problems, it is still necessary to propose algorithms to effectively solve the large-scale planning problems that appear in real-world enterprises to ensure that the optimal or near-optimal solutions are robust when different variables interact.

In order to bridge the gaps in the literature, a framework is proposed for dealing with the formulation of replenishment, production and delivery plans, and for proposing solution algorithms to efficiently deal with such complexity. This framework is used to: (i) identify the type of planning problem to be represented by the modeller, and the associated objective function; (ii) generate a range of input data to be potentially used for modelling the desired plan by considering the defined objective function; (iii) provide a MILP skeleton that consists in an open mathematical modelling language. This skeleton is characterised by being versatile enough to be applied to any studied plan object based on modellers' requirements; (iv) select the algorithm that is most likely to solve MILP; (v) build a standard structure and implement the previously selected algorithm.

3.1   *Methodology to Formulate Mathematical Models*

Mathematical programming models spend extremely long computing times and, therefore, it is in modellers' interest to build quickly formulated models. The proposed methodology follows a set of steps (Table 1) to formulate mathematical programming models, which are to be potentially applied to develop any planning model regardless of its nature. The objectives, input data [21] and output data are classified per plan type S [22], M [19], D [23], SM [22], MD [23], SMD [19] (see Table 2).

Table 1 Methodology steps to formulate a mathematical model in the planning context

Step 1 plan type	Determine the type of plan to be modelled [5, 18]: (i) Source (S), replenishment plans; (ii) Make (M), production plans; (iii) Deliver (D), transport plans. It is also interesting for modellers to solve a combined type of plans, in which a collaborative perspective of the planning problem is addressed [19] (i) Source and Make (SM); (ii) Make and Deliver (MD); (iii) Source Make and Deliver (SMD)
Step 2 plan subtype and horizon	Identify the plan subtype to be modelled. When defining a plan, we may think that the plan subtype implicitly concerns the planning horizon. Sometimes this situation happens, e.g. when the planning problem to be modelled is a scheduling plan, the horizon covers only a few weeks; or the horizon in aggregate production plans is set at 1 year. In this step, apart from indicating the plan subtype, the model designer has to identify the time horizon and the periods into which the horizon is divided. Periods allow the identification of dynamic changes, i.e. demand variation, which occur in the planning horizon. Plan types and plan subtypes are defined by Andres and Poler [20], and a summary of them is presented below: • **Source**: Inventory planning; Procurement planning; Material requirements planning; Replenishment planning • **Make**: Finished good inventory planning; Production planning; Production Scheduling; Production sequencing • **Deliver**: Demand Planning; Distribution planning; Order-Promising; Transport planning • **Source and Make**: Materials requirement planning & Production Planning; Inventory planning & Production planning • **Make and Deliver**: Production planning & Distribution planning, Production planning & Transport planning • **Source Make and Deliver**: Inventory planning & Production planning & Distribution planning; Replenishment planning & Production planning & Distribution planning

(continued)

Table 1 (continued)

Step 3 objectives	Select the objectives to be optimised according to the object plan type and plan subtype to be modelled. The objective function is the result of mathematically representing a planning goal to be used in decision analyses, operations research or optimisation studies. The commonest objective functions aim to minimise the expected benefit or the utilisation ratio. However, the objective functions proposed in the framework are not only limited to these two objective types, but other objective functions could become relevant in some planning problems; e.g. (i) maximise profit; (ii) minimise costs; (iii) maximise total production in units; (iv) minimise production time; (v) maximise the market share for all or some products; (vi) maximise total sales in units or monetary units; (vii) minimise production pattern changes; (viii) minimise the use of a limited material components or products; (ix) minimise number of employees; (x) maximise customer satisfaction. To minimise costs, it is important to set appropriate restrictions because sometimes minimising costs means doing nothing. We must also properly distinguish fixed costs and variable costs. To maximise profits, modellers must bear in mind that they can be made over time. Incorporating the time concept into the evaluation of profit can be done in many ways, among which the Net Present Value stands out
Step 4 input data	According to the selected objectives, a set of representative input data is proposed by the framework. The modeller has to select the input data existing in the enterprise, for which the selected plan is modelled. The input data comprises the parameters of the mathematical programming model. Parameters are beyond the control of the decision-maker and are imposed by the external environment. The parameters represent those factors that affect the decision but are not controllable directly (such as prices, costs, demand, and so forth). In deterministic mathematical programming models, all the parameters are assumed to take fixed, known values, where estimates are provided via point forecasts. The impact of this assumption can be tested by means of sensitivity analysis. Examples of some of the parameters associated with a production planning problem are: product demands, finished product prices and costs, productivity of the manufacturing process, and manpower availability [17]. Knowing the types of data available allows establishing the sets and, with them, the indexes. The representation of data sets, using symbols with subscripts, will allow the conceptualisation of the problem

(continued)

Table 1 (continued)

Step 5 restrictions	Considering inputs, a set of standard restrictions is selected. Mathematical programming restrictions express relations between variables and take the formulation of a linear combination of variables limited by a certain value. Restrictions can be classified according to: (i) capacity restrictions; the production of a set of products is limited because some of the resources used in their manufacturing are limited (machines, labour, schedule); (ii) raw material availability; production of a set of products is limited according to the amount of raw material available; (iii) limitations in market demand; the production of a product is limited based on the estimated sale; (iv) continuity restrictions or material balance; during multiperiod programming, the products that remain at the end of one period are those that exist at the beginning of the next one; (v) quality stipulations; when mixing products, restrictions can be set based on the quality characteristics of the mixture and raw materials; (vi) logical-type relations
Step 6 output data	Given the selected objectives and the identified input data, a list of output data is proposed in the framework. Modellers must select the output data, which is interesting for the enterprise. The output data consist of the set of decision variables to be solved in the mathematical programming model. The decision variables are those factors under the decision maker's control, and result in the answers that decision makers seek. From Step 3, the variables that configure the objective function are defined. Here the intention is to define values for these variables so that the best assessment of the objective function is made, while all the restrictions are met. Some examples when modelling production planning models are: (i) the amount to be manufactured of each product during each time period; (ii) the amount of inventory that accumulates during each time period; (iii) regular hours and overtime labour during each time period
Step 7 model skeleton	The framework proposes a mathematical model MILP skeleton of the planning problem to be modelled. The proposed model skeleton provides a compact realistic model in which different variables implicitly appear. The skeleton uses acronyms to designate variables and constraints so that the results can then be interpreted more easily. Although less compact models, such as those proposed by this framework, require a longer resolution time, this time is compensated by the length of time to be invested in interpretating the solution
Step 8 modellers' adjustment	Modellers or enterprise planners tune the proposed mathematical model MILP skeleton by considering the enterprise's specific characteristics. This analysis is already leading to a better understanding of the problem. The adjustment and validation process are repeated until the model sufficiently and accurately represents reality. This step is very useful for understanding the modelled reality itself

Table 2 Objectives, input data [21] and output data classified per plan type S [22], M [19], D [23], SM [22], MD [23], SMD [19]

	Plan type	Nomenclature
Objectives	S	Inventory cost minimisation, Profit maximisation, Idle time minimisation, Backorder's minimisation
	M	Production cost minimisation, Profit maximisation, Setup minimisation
	D	Transport cost minimisation, Sales maximisation, Inventory minimisation, Backorder's minimisation, Service level maximisation
	SM	Inventory cost minimisation, Profit maximisation, Idle time minimisation, Backorder's minimisation, Production cost minimisation, Transport cost minimisation
	MD	Transport cost minimisation, Sales maximisation, Inventory minimisation, Backorder's minimisation, Service level maximisation
	SMD	Production cost minimisation, Profit maximisation, Setup minimisation, Transport/distribution cost minimisation
Input data	S	Demand, Inventory, Capacity, Production Time, Setup, Bill of Materials (BOM), Supply Lead time, Supplier prices
	M	Demand, Inventory, Capacity, Production Time, Set-up and BOM
	D	Demand, Inventory, Capacity, Production Time, Transport/Distribution Cost, Backorders, Supply Lead time, Supplier Prices
	SM	Demand, Inventory, Capacity, Production Time, Setup, BOM, Supply Lead time, Supplier Prices
	MD	Demand, Inventory, Capacity, Production Time, Transport/Distribution Cost, Backorders, Supply Lead time, Supplier Prices
	SMD	Demand, Inventory, Capacity, Production Time, Setup, BOM, Transport batch minimum Transport Capacity
Output data	S	Components to purchase, Backorder's, Inventory, Delivery time
	M	Products to produce, Backorder's, Machine assignation and Overtime
	D	Transport cost, Backorder's, Inventory, Delivery time, Total cost, Product to transport
	SM	Components to purchase, Backorder's, Inventory, Delivery time, Products to produce
	MD	Transport cost, Backorder's, Inventory, Delivery time, Total cost, Product to transport
	SMD	Products to produce, Backorder's, Machine assignation, Overtime, Raw material to purchase, Product quantity to transport

3.2 Identifying Solver Algorithms

The computational cost for solving large-scale industrial problems is still excessive today. Some general solution procedures are available, can be purchased on the market and are capable of solving increasingly complicated problems in appropriate times. In practice, however, it may be more cost-effective to design the solution procedure. Therefore, methods for designing problem-solving procedures are already modelled and are addressed in this section of the paper.

Although modellers can programme algorithms to solve mathematical planning problems, it is worth noting that occasionally using commercial software is more efficient than any individual implementation, such as CPLEX and Gurobi. Apart from optimisation software, it is necessary to have interface software to not only access and collect data, and to also structure and introduce a problem into a model-shaped package. Indeed, different packages provide high-level languages for mathematical programming, e.g. MPL modelling language from the Maximal Software, JUMP (Julia) or Pyomo (Python).

An exact algorithm ensures obtaining the best possible solution, the optimal one, by exploring the entire solution space. Nevertheless, the methods commonly used to solve problems are of a heuristic or metaheuristic type. Heuristics, metaheuristic and matheuristic algorithms are capable of generating approximate solutions for the problem and come as close as possible to the optimum one but may fail while making attempts. Being able to design a good heuristic, metaheuristic or matheuristic algorithm requires knowledge of the problem, which can lead to other improvements. This section of the proposed framework helps modellers to identify the most appropriate solver algorithm according to the identified plan type and plan subtype (steps 1 and 2 in the methodology). The algorithms proposed by the framework consists in a procedure that allows a solution for the selected specific planning problem to be found (see Table 3).

3.3   Identifying Appropriate Algorithms

Algorithms consist of a systematic procedure that moves from one decision point to another to solve a category of problems. The Simplex algorithm is, for example, used to solve LP problems. The algorithms proposed in this part of the framework always meet one of three conditions: (i) there is no feasible solution; (ii) there is an optimal solution; (iii) the objective function is not limited to the feasible region. Moreover, the algorithms need: (i) procedure initialisation; (ii) a stopping criterion to denote when a solution is reached; (iii) an improvement method to move from an area of solution where there is no solution (relative minimum or maximum) to a better area to achieve the optimal or near-optimal solution. Algorithms can be classified according to the proximity to the optimum and the calculation mechanism [20]: (i) optimiser (AO), an algorithm that follows a systematic procedure that ensures achieving the optimum solution. Nevertheless, for some classes of problems, the time required to find the optimal solution is unacceptable. Algorithms that enable good solutions to be found are needed, including: (ii) heuristic (AH), an algorithm that employs an *ad hoc* procedure, but does not guarantee reaching the optimal solution, rather a near-optimal or sufficient one for immediate objectives; (iii) metaheuristic (AM), which is a higher-level procedure followed to select a heuristic (partial search algorithm) to obtain a sufficiently good solution. AM includes random searches that

Table 3 Solver algorithms used per plan type S [22], M [19], D [23], SM [22], MD [23], SMD [19]

Algorithm type	S	M	D	SM	MD	SMD
AMT/ Collaborative Agents		X			X	
AH/ Campbell–Dudek Algorithm		X				
AH/ Local improvement procedure		X				
AH/ Multi-Objective Master Planning Algorithm		X				
AH/ Primal–Dual-Based Heuristic		X			X	
AH/ Variable Neighbourhood Search						
AH/ Decomposition & Aggregation			X			
AH/ Greedy	X	X				
AH/ Greedy			X			
AH/ Lagrangian					X	X
AM/ Genetic Algorithm	X	X	X		X	
AM/ Iterated Local Search	X					
AM/ Simulated Annealing	X					
AM/ Tabu Search		X				
AM/ Tabu Search Grabowski and Wodecki		X				
AO/ Decomposition strategy		X				
AO/ Fuzzy Programming			X			
AO/ Lomnicki		X				
AO/ Solution procedure of model P*		X				
AO/ Strategic-operational optimisation solution algorithm		X				X
AO/ Branch and Bound	X		X			
AO/ Branch and Bound		X	X			
AO/ Dynamic Programming	X	X	X			
AO/ Lompen Algorithm		X				
AO/ Simplex	X	X	X	X	X	X

facilitate achieving several solutions (without ensuring the optimum) and needs a termination rule; (iv) matheuristic (AMT), which is a procedure that consists in the interoperation of metaheuristic and optimisation techniques [3]. Matheuristic can find near-optimal solutions (or sufficiently good ones) more quickly than some optimisation procedures. The reviewed papers indicate the use of each algorithm according to the plan type (Table 3).

4   Case Study

The case study is generated for a particular planning problem at the operational decision-making level, namely the scheduling plan of the second-tier supplier in the automotive supply chain as part of the "Zero-Defect Manufacturing Platform" (ZDMP) H2020 Project. The framework herein proposed is applied by the authors using realistic data. The plan type is determined by the Make classification of SCOR. The scheduling plan deals with the start and due dates of individual products, and also with machine assignments. It involves allocating finite resources to meet demand requirements by contemplating constraints like capacity, precedence and start and due dates, and identifying the quantity of products to be produced during a certain period [19].

In order to obtain a representative amount of parameters and variables to create the scheduling plan's skeleton, a literature review is done in the scheduling context [21, 24–28]. The review process allowed us to identify a set of objectives, input data and output data, which are classified according to their nature (see Table 4): (i) capacity: referring to the amount of resources the enterprise owns for planning, e.g. number of workers, time, space, machines, monetary units, etc.; (ii) inventory: concerning the properties of the products in the warehouse; (iii) product: applied to the features related to raw materials and finished products; (iv) production: characterises the processes and methods used to transform raw materials, semifinished goods and subassemblies; (v) resources: seen as the productive factor required to perform an activity to obtain final products; (vi) sequence: contemplates the dependence and precedence of materials, products or resources; (vii) time: related to the unit of measurement used to categorise length of time; (viii) transport: considers the aspects related to moving products from one place to another. The main indices applied in the reviewed works are: set of products (finished goods, raw materials); set of finished goods; set of periods.

The proposed framework also provides a set of common constrains that characterise the scheduling plan, including: (i) inventory balance equations for finished goods and raw materials; (ii) inventory capacity limitation; (iii) production capacity limits; (iv) production sequence determination; (v) the product for which the machine is setup; (vi) only one product can be setup at the end of each period; (vii) elimination of subtours when more than one product is produced during a single period. The same product cannot be produced as both first and last during a period; finally (viii) the binary and non-negativity properties for the decision variables are to be included.

The framework shows the model designer all the objectives, input data, output data and constraints to select the parameters, variables and restrictions that apply to the enterprise's scheduling plan. According to the selected elements, the framework generates the mathematical model skeleton (MILP). Finally, the modeller reviews the proposed MILP and makes final adjustments.

The framework identifies the most appropriate solver algorithm to solve the end version of MILP. The application of a solver algorithm such as a heuristic algorithm allows large-sized problems to be solved, which involves a vast amount of data, and

Table 4 Scheduling plan: objectives, input data and output data

Objetive	Inventory	Cost (or units) below safety stock minimisation; Inventory cost minimisation
	Product	Backorder minimisation (quantity or cost); Value of products maximisation; Raw Material cost minimisation; Utilities cost minimisation
	Production	Production cost minimisation; Profit maximisation; Overtime minimisation
	Sequence	Sequencing cost minimisation; Setup cost/time minimisation
	Time	Makespan minimisation
Input data	Capacity	Maximum Inventory; Minimum inventory; Utility capacity; Production capacity
	Inventory	Inventory cost; Safety stock shortage cost; In factory products Inventory; In Factory raw materials inventory; Set of units suitable for temporarily storing; Inventory capacity; Scheduled receptions
	Product	Production batch minimum; Bill of materials; Demand; Items to be produced; Backorder's cost; Material cost; Product sequence; Sequencing rules for option o; Product sequence permutation; Production batch target; Production cost; Infeasible set of operations sequencing; Delivery priority
	Resource	Cost of order processing jobs; Job Profit; Number of jobs; Normal machine capacity; Tasks required in machines; Set of processing tasks that can be performed on a machine; Assigning tasks to machines; Machine tools number; Normal machine cost; Utility cost
	Sequence	Setup cost; Setup times dependent on sequencing
	Time	Backorders maximum delay allowed; Due date; Horizon; Lower and upper bound on the allowable end time of an outage product; Period; Slots; Processing time; Production time; Overtime cost
Output data	Inventory	Inventory level of the product at the end of the period
	Product	Product quantity to produce; Production batch; Production time of the product; Product produced during a period; Product produced first in the period; Products produced last in the period
	Resource	Allocating tasks to a machine at the beginning at time; Assigning a machine
	Sequence	Orders sequence; Remaining elements to be sequenced; Setup from product i to product j during a period; Variables to eliminate subtours
	Time	Delivery times; Due date; Product lateness; Overtime

the complete enterprise scheduling problem is considered. This means scheduling all the products manufactured by the enterprise using each involved resource and a real time horizon.

5 Conclusions

This paper identifies the gap identified in the literature about the automatic formulation of mathematical models and solver algorithms to solve large-sized planning problems. As far as we know, the papers that propose guidelines and tools to mathematically formulate planning problems are limited to model formulation, and do not take into account enterprises' real needs, e.g. formulating models applicable to solve large-sized enterprise plans. The main contribution of this paper led to the proposal of a complete framework, which allows planning processes to be modelled by considering not only an intra-enterprise perspective that involves replenishment, production and delivery plans, but also collaborative scenarios in which supply chain plans are jointly solved. The framework also focuses on identifying solver algorithms, which can manage large amounts of data and allow planning models to be solved in a computationally efficient manner.

The advantages of using mathematical models derive from the clear conceptualisation of the industrial planning process to be modelled. However, the modeller must know that there are times when the mathematical formulation is limited by having to generate artificial constraints to model restrictions that can be easily modelled with a heuristic algorithm. Moreover if the problem's behaviour is nonlinear, applying a LP model can only model an approximation to reality, and more artificial restrictions should be created. The identified limitations enabled the authors to identify future research lines that lead to the framework being extended so as to not force users having to face developing a mathematical model. In this way, the modeller can directly generate a heuristic or metaheuristic to model the planning problem and solve it. A second future research line is about examining in more depth the part of the framework employed to identify and formulate solver algorithms. Here the first action is to focus on generating the solver algorithm. The second action goes further and permits the authors to propose general simple metaheuristic procedures to solve large-sized planning problems in short times with fewer computational resources.

Acknowledgements The research leading to these results was partly supported by the European Commission with Grant Agreement No. 825631 "Zero Defects Manufacturing Platform" (ZDMP) (www.zdmp.eu); The present research was supported by the Conselleria de Educación, Investigación, Cultura y Deporte (Generalitat Valenciana) to hire predoctoral research staff with Grant ACIF/2018/170, and by European Social Funds with Grant "Operational Programme FSE 2014–2020" of the Valencian Community (Spain).

References

1. Christou T. I. (2012). *Quantitative methods in supply chain management: Models and algorithms*. Springer-Verlag.
2. Campuzano F. & Mula J. (2011). *Supply chain simulation: A system dynamics approach for improving performance*. Springer.

3. Boschetti M. A., Maniezzo V., Roffilli M., & Bolufé Röhler A. (2009) Matheuristics: Optimization, simulation and control. Lecture Notes Computer Science (including subseries lecture notes artificial intelligent lecture notes bioinformatics), vol 5818. LNCS (pp. 171–177).
4. Andres, B., & Poler, R. (2016). A decision support system for the collaborative selection of strategies in enterprise networks. *Decision Support Systems, 91*, 113–123.
5. Supply chain council (2012). Supply chain operations reference model (SCOR).
6. APICS (2017). 'SCOR framework', Supply chain operations reference model (SCOR).
7. Pidd, M. (1996). *Tools for thinking: Modelling in management science.* Wiley.
8. Hackman, S. T., & Leachman, R. C. (1989). A general framework for modeling production. *Management Science, 35*(4), 478–495.
9. Krishnan, R. (1990). A logic modeling language for automated model construction. *Decision Support Systems, 6*(2), 123–152.
10. Krishnan, R. (1991). PDM: A knowledge-based tool for model construction. *Decision Support Systems, 7*(4), 301–314.
11. Shapiro, J. R. (2001). *Modeling the supply chain.* HAPIRO Duxbury Press.
12. Hernández, J. E., Mula, J., Ferriols, F. J., & Poler, R. (2008). A conceptual model for the production and transport planning process: An application to the automobile sector. *Computers in Industry, 59*(8), 842–852.
13. Hernandez, J. E., Mula, J., & Ferriols, F. J. (2008). A reference model for conceptual modelling of production planning processes. *Production Planning Control, 19*(8), 725–734.
14. Hashimoto H. & Kubo M. (2016) Supply chain optimization: A survey—models, algorithms, decision support systems, and applications. *10*(1), 9–18.
15. Mula, J., Peidro, D., Díaz-Madroñero, M., & Vicens, E. (2010). Mathematical programming models for supply chain production and transport planning. *European Journal of Operational Research, 204*(3), 377–390.
16. Prasad, T., Saleem, A., Srinivas, K., & Srinivas, C. (2017). Supply chain management—Modeling and algorthims: A review. *International Journal of Mechanical Engineering, 8*(3), 191–197.
17. Alemany, M. M. E., Alarcón, F., Lario, F. C., & Boj, J. J. (2011). An application to support the temporal and spatial distributed decision-making process in supply chain collaborative planning. *Computers in Industry, 62*(5), 519–540.
18. Schroeder, W., & Beale, E. M. L. (1968). Mathematical programming in practice. *Operations Research, 19*(4), 487.
19. Andres B., Poler R., Saari L., Arana J., Benaches J. V., & Salazar J. (2018) Optimization models to support decision-making in collaborative networks: A review. In *Closing the Gap Between Practice and Research in Industrial Engineering*, Lecture Notes in Management and Industrial Engineering (pp. 249–258).
20. Andres, B., & Poler, R. (2016). Models, guidelines and tools for the integration of collaborative processes in non-hierarchical manufacturing networks: A review. *International Journal of Computer Integrated Manufacturing, 2*(29), 166–201.
21. Andres B., Sanchis R., Poler R., & Saari L. (2017). A proposal of standardised data model for cloud manufacturing collaborative networks. vol 506.
22. Orbegozo A., Andres B., Mula J., Lauras M., Monteiro C. & Malheiro M. (2016). An overview of optimization models for integrated replenishment and producction planning decisions. In *Building bridges between researchers and practitioners.* Book of Abstracts of the International Joint Conference CIO-ICIEOM-IISE-AIM (IJC2016) (p. 68).
23. Andres, B., Sanchis, R., Lamothe, J., Saari, L., & Hauser, F. (2017). Integrated production-distribution planning optimization models: A review in collaborative networks context. *International Journal of Production Management and Engineering, 5*(1), 31–38.
24. Reyes Y., Mula J., Diaz-Madronero M., & Gutierrez E. (2017) Master production scheduling based on integer linear programming for a chemical company. *Revista de Métodos Cuantitativos para la Economía y la Empresa, 24*, 147–169.
25. Andres B., Sanchis R., Poler R., Mula J., & Díaz-Madroñero M. (2017) A MILP for multi-machine injection moulding sequencing in the scope of C2NET Project. *International Journal of Production Management and Engineering*, vol. In press (pp. 29–36).

26. Diaz-Madroñero M., Mula J., Andres B., Poler R., & Sanchis S. (2018). Capacitated lot-sizing and scheduling problem for second-tier suppliers in the automotive sector. Lecture Notes Management Industrial Engineering (pp. 121–129).
27. Andres B., Poler R., Sanchis R., Mula J., Díaz-Madroñero M. (2019) Interoperable algorithms for its implementation in a cloud collaborative manufacturing platform. In *Enterprise Interoperability VIII, Proceedings of the I-ESA Conferences, Enterprise*, vol. 9, P. K., T. KD., K. T., and P. R., (Eds.), Springer International Publishing (pp. 1–443).
28. Martín A. G., Díaz-Madroñero M., & Mula J. (2019) Master production schedule using robust optimization approaches in an automobile second-tier supplier. *Central European Journal of Operations Research*.

Introduction to a Physics-Based Theory to Manage Risks and Opportunities in Supply Chains

Thibaut Cerabona, Frederick Benaben, Louis Faugère, Matthieu Lauras, Jean-Philippe Gitto, and Benoit Montreuil

Abstract Currently, the management of risks and opportunities is highly depending on the ability of managers: to analyze complex situations (with a lack of common vision and decision on the whole supply network), to mobilize their experience and their knowledge. The goal of this article is to introduce a new vision for collaborative network management, especially dedicated to supply chain management. It deals with an innovative and original approach for supply chain management, based on physical principles. With that theory, risks and opportunities can be seen as forces pushing and pulling a system according to its key performance indicators (KPIs).

Keywords Risk and opportunity management · Supply chain · Physics-based theory

1 Introduction

Managing a supply chain involves shaping and pursuing objectives. These can be represented by KPIs. Managers tend to like measurable objectives, even though sometimes they are not looking at the "good" data. Within this article, we consider that the majority of objectives can be evaluated through formal KPIs. Managing a supply chain is trying to bring its KPIs to predefined targeted values.

The evolution of the KPIs is due to the occurrence of potentialities, when they become actualities. As defined in [1], a risk is the negative deviation from the expected value of a certain performance measure (a KPI for example), from which result negative consequences for the organization. Symmetrically, an opportunity is

T. Cerabona (✉) · F. Benaben · M. Lauras
Centre Génie Industriel, IMT Mines Albi, Université de Toulouse, Albi, France
e-mail: thibaut.cerabona@mines-albi.fr

L. Faugère · B. Montreuil
ISyE School, Physical Internet Center, Supply Chain and Logistics Institute, Georgia Tech, Atlanta, USA

J.-P. Gitto
Scalian, Toulouse, France

a potentiality whose consequences are positive with regard to the target performance indicators.

Besides, a supply chain is a collaborative network, dealing with risks and opportunities for the whole network requires a global vision: thus, the suggested approach could contribute to the holistic vision of the supply chain and to interoperability of the network.

This article claims that (i) the identification of objectives and (ii) the support for decision-making are essential to the management of an organization. These decisions provide a chance to seize opportunities or escape risks in order to achieve the target values of the KPIs.

This article answers that following question: "*how to define and control the trajectory of the overall performance of a supply chain with its risks and opportunities represented as forces?*". This article is structured as follows: Sect. 2 provides a state of the art regarding risks and opportunities management in supply chain. Section 3 introduces this theory. Section 4 illustrates that theory with a supply chain management use-case. Finally, the last section concludes with some perspectives.

2   State of the Art Regarding Risks and Opportunities Management in Supply Chain

Supply chain is earmarked by predictable or unexpected events that threaten the reach its performance objectives [2]. So its management implies to deal with risks and opportunities. As discussed in [3], the concepts of risk and opportunity are in fact very close to each other. These concepts are symmetric: pull or push the considered system in relation to its aims. Indeed, generally speaking, opportunity is the opposite of risk but both impact the location of a system with regards to its KPIs (making it closer or farer to the target values). In the following, the article focuses on the concept of risk, only because the field of risk management and detection is considerably more developed and studied.

In this section, the concept of risk and opportunity will be explored from the literature in order to provide guidelines for characterizing risks and opportunities. First of all, the risk management processes are decomposed in four steps as described in [4]:

- Risk Identification (detection of potentialities by studying an organization and its environment).
- Risk Assessment (evaluation of the impact of risk on the system, it can be divided in two parts: qualification and quantification).
- Risk Mitigation (risks responses strategies such as: acceptance, avoidance, sharing).
- Risk Monitoring (monitor the status of previously identified risks).

In [5], risk is a combination of the impact on the organization and its probability of occurrence. The probability is a way to balance the impact seen as the consequences on the KPIs. This is a very used two dimensions' framework for risk analysis as presented on Fig. 1 (including also opportunity):

Even if this idea of potentiality (risk and opportunity) being considered as the combination of impact and probability is very well recognized and used, this paper claims that, as discussed in [6], this vision can be refined according to the following three basic components:

- The danger(s) or a driver(s) (induct the risk)
- The event(s) (including the probability of occurrence)
- The consequence(s) (the real impact(s) of the occurrence of the risk).

In [7], a causal vision of the cascading risk chain is based on a very similar structure: *danger-risk-consequence*, so-called the DRC chain. Besides, a very interesting aspect of that vision is that it can easily be extended to describe the opportunity by including the notion of favorable condition (i.e., the positive version of a *danger*).

Thus, that cascading effect chain can be generalized as illustrated in Fig. 2 (from [8]).

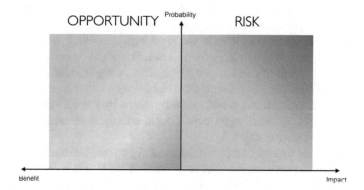

Fig. 1 Two-dimensional framework for risk and opportunity analysis

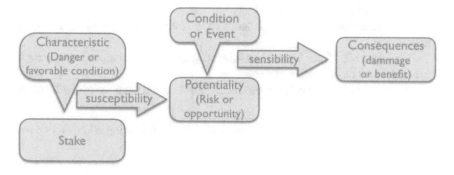

Fig. 2 Causal chain for potentialities

The existing research results on the field of risk management can be symmetrically extended to the question of opportunity management. At the end of the day, both together can be considered as potentiality management. However, the existing approaches on risk identification are essentially based on literature review and other experience-based methods [9, 10]). Therefore, it is possible to suggest the following two points:

- The impact/probability model is quite simple and cannot be exploited to model cascading effect.
- There is no automated tool for potentiality identification based on the use of data.

3   A Physics-Based Theory

Now dealing with instability is a norm, as stated by [11, 12]. But it is a new perspective of management and opens the door to new theories [8]. It is the purpose of this paper to introduce a new and original approach for supply chain management. This approach applies physical principles for supporting decision-making processes to control a supply chain's trajectory.

With that theory, risks and opportunities are modeled by forces pushing or pulling the considered supply chain within its KPI framework (a risk can be seen as a force pushing the supply chain away from its target performance values, while an opportunity would bring it closer to its target performance values). For example, a perspective like "solid mechanics" can be done, especially to study the managed or inflicted deformations of a supply chain due to the faced forces.

In addition to their direction and intensity (given by the KPI framework), the obtained forces are different types which could constitute interesting patterns of impact matrix and vectors [8].

Indeed, there are two major types of risks and opportunities based on quantifiable and measurable dimensions: external or internal [13, 14]. So, we can define four different force natures to model them (the first two for the external part and the last two for the internal part), as explained in [8]:

- External field force: is a force induced by an external characteristic (new taxes, hurricane, etc.).
- Collaboration force: is a force induced by a partnership (a supplier, a subcontractor, etc.).
- Internal force: is a force induced by internal decisions (continuous improvement, buying a new machine, poor forecast, etc.).
- Gravity force: is a force induced by unavoidable internal weights (operating costs, etc.).

These four different types of forces are based on the following criteria as we can see in Fig. 3:

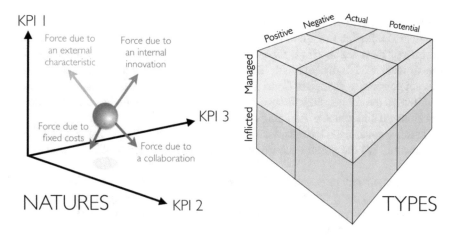

Fig. 3 Complementary characteristics of forces

- Is the force real or potential?
- Is the force positive or negative with regards to the considered KPIs, i.e., does it model an opportunity or a risk?
- Is it a force that the considered system is able to manage? Or, is it inflicted on it?

This framework of KPIs (Fig. 3) allows to locate the considered supply chain regarding its KPIs and to define its network of forces (each force reflects the probable consequences of each identified risk or opportunity). For a supply chain, the dimensions of the framework could be: cost, quality, time, but also less quantitative dimensions like reactivity, flexibility, robustness, stability, or resilience ([15] key dimensions of today's performance).

This decision-making space can be used to determine the target zone, a part of this space reflecting the target of the considered supply chain in terms of KPIs. By looking intensity of these identified forces, one can study how to select the best combination of potentialities and the required effort to reach the target zone.

4 An Illustrative Example in the Field of Supply Chain Management

Please note that: This use-case is not based on real facts. Its purpose is to illustrate the theory presented in Sect. 3.

The considered system is the supply chain of a spring for truck manufacturer. This plant can catch a big opportunity: to manufacture the further generation of leaf springs for its biggest customer. The plan manager does not want to miss this vital opportunity. It could be the future flagship product for his plant, so he puts a big pressure on the supply chain manager. Some trials are realized and were conclusive. The factory starts to produce the first small series, but a quality problem was detected. This problem is linked to the use of rubber silent blocks. Indeed, during the manufacture process of these leaf springs, rubber bushings must be fit into their eyelets and are tore or weakened in irregular rate. This problem is due to this new eyelet design, and it is not possible to change it. Consequently, the supply chain has to make a decision: continue to use rubber silent blocks (i.e., continue to work with the current supplier) or use metal silent blocks (i.e., find another supplier).

In this use-case, the two characteristics studied will be: (i) the raising sales of this product and (ii) the possible choice of a new supplier which produces metal silent blocks.

The considered performance framework is composed of the following three dimensions: (i) cost, (ii) delay and (iii) quality. Clearly, management's goal is to minimize these three considered KPIs (note that the concept of time is included in the concept of trajectory):

- Financial rate:

$$\text{Financial rate} = \frac{\text{Number of parts sold.}(\text{Manufacturing costs} + \text{Investments})}{\text{Sales price}}$$

- Delay rate:

$$\text{Delay rate} = \frac{\text{Number of a customer's orders with a delay in delivery bigger than one day}}{\text{Number of its orders}}$$

- Scrap rate:

$$\text{Scrap rate} = \frac{\text{Number of scraps}}{\text{Number of used silent blocks}}$$

At the end of the first production orders, this is the position of our supply chain in its framework of KPIs. A potentiality is to choose a new supplier, modeled by these initial matrix and vectors:

$$F_{\text{new supplier}} = \begin{bmatrix} 1.5 & 0 & 0 \\ 0 & 1 & 0 \\ 0 & 0 & 0.9 \end{bmatrix} \cdot E + \begin{bmatrix} 2 \\ 0 \\ -0.5 \end{bmatrix} \text{ with } E = \begin{bmatrix} E_{\$} \\ E_{\text{Delay}} \\ E_{\text{Quality}} \end{bmatrix}$$

The trends observed by this expression of this potential force can be explained by:

- 1.5: metal silent blocks are more expensive
- 1: any change on the process time
- 0.9: new silent blocks improve the quality of each spring
- 2: the plant invests in an adapted machine for these news bushings

- 0: any change on the delivery time
- −0.5: new silent blocks reduce the number of scraps.

The scales are normalized for clarity reason. As illustrated in Fig. 4, the same supplier forces (i.e., the force which represents the raising sales of this new product with a rubber bushing) and the new supplier force (i.e., the force which represents the raising sales of this new product with a metal bushing) are not pushing the supply chain in the same direction. The new supplier force is a potential collaborative force. If the manager chooses to continue with the current supplier, the supply chain may move in the framework and new forces will be associated to that new position (see figure). The following matrix and vectors are those of this new position:

$$
F_{\text{same supplier}} = \begin{bmatrix} 2 & 0 & 0 \\ 0 & 1.5 & 0 \\ 0 & 0 & 4 \end{bmatrix} \cdot E + \begin{bmatrix} 2 \\ 1 \\ 3 \end{bmatrix}
$$

The trends observed by this expression of this force can be explained by:

- 2: the manufacturing costs increase due these high rate of scrap
- 1.5: the process time increases sharply with all these scraps
- 4: the scraps increase
- 2: the plant invests to improve the accuracy of the machine
- 1: as the process time and quantities ordered increase, there is a risk of delay
- 3: the overall quality of the products decreases, risks tarnishing its image.

Fig. 4 Initial situation and forces

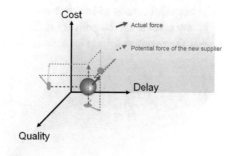

Fig. 5 Possible position and forces of the supply chain if the company maintains the current supplier

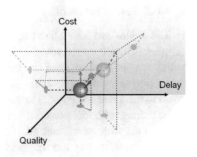

In Fig. 5, the light red sphere shows the further possible position of our supply chain, pushed only the raising sales of this product with the current supplier force. As a consequence of that new position and the previous force, a potential force has been calculated (green vectors). It is important to note that the new vector associated with the choice of the new supplier has not been calculated.

If the decision to choose a new supplier is taken, the supply chain may move in the framework and new forces will be associated with that new position modeled by these new matrix and vectors (see Fig. 6):

$$F_{\text{new supplier}} = \begin{bmatrix} 3 & 0 & 0 \\ 0 & 1 & 0 \\ 0 & 0 & 0.7 \end{bmatrix} \cdot E + \begin{bmatrix} 2 \\ 0 \\ -1 \end{bmatrix}$$

The trends observed by this expression of this force can be explained by:

- 3: metal silent blocks are more expensive and sales volumes increase
- 1: any change on the process time
- 0.7: the parts are more robust
- 2: the plant invests to improve the accuracy of the machine
- 0: any change on the delivery time
- −1: the recent machine is more accurate and the scraps go down.

Fig. 6 Possible position and forces of the supply chain if the company takes the new supplier

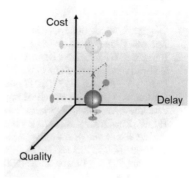

In Fig. 6, the new supplier force pushes the supply chain to a possible position represented by the orange sphere.

Finally, the best decision for the supply chain manager would be to choose a new supplier. For this use-case, this decision is relatively simple. Figure 5 allows to easily understanding that if the manager does not change anything, the forces acting on its supply chain will push it to very bad results.

5   Conclusion and Perspectives

Dealing with instability is a new management perspective [8]. In order to adapt to the forces in presence, the mechanics of solids will be taken into consideration, especially to study the chosen or inflicted deformations of an organization. The theory presented in this paper opens the door to an innovative vision for management and decision-making. The aim being to develop an intuitive decision tool to support inter-enterprise collaboration. With that tool, managers could see the impacts of each potentiality on the collaboration performance and help interoperability. In return, interoperability contributes to one type of forces (collaboration force) by favoring the exchange of information and thus helping to improve the quality of the relationship.

The following points are a set of identified tasks to make this theory operational:

- The real-time calculation of trajectories.
- The study of accessible KPI space areas and efforts to join these areas.
- The visualization, being able to observe the supply chain within the framework of its KPIs. This last aspect is very important, because frameworks usually include more than three dimensions and are already hard to manage. Virtual reality is considered a potent way for supporting such visualization.

References

1. Wagner, S. M., & Bode, C. (2006). An empirical investigation into supply chain vulnerability. *Journal of Purchasing and Supply Management, 13*, 301–312.
2. Baryannis, G., Validi, S., Dani, S., & Antoniou, G. (2019). Supply chain risk management and artificial intelligence: State of the art and future research directions. *International Journal of Production Research, 57*(7), 2179–2202. https://doi.org/10.1080/00207543.2018.1530476
3. Olsson, R. (2007). In search of opportunity management: Is the risk management process enough? *International Journal of Project Management, 25*(8), 745–752.
4. Ho, W., Zheng, T., Yildiz, H., & Talluri, S. (2015). Supply chain risk management: A literature review. *International Journal of Production Research, 53*(16), 5031–5069. https://doi.org/10.1080/00207543.2015.1030467
5. Edwards, P. J., & Bowen, P. A. (2005). *Risk management in project organisations.* Elsevier, Oxford, UK.
6. Benaben, F., Montreuil, B., Gou, J., Li, J., Lauras, M., Koura, I., & Mu, W. (2019). A tentative framework for risk and opportunity detection in a collaborative environment based on data interpretation. In *Proceedings of 52nd HICSS'19*, Hawaii, USA.
7. Benaben, F., Barthe-Delanoë, A. -M., Lauras, M., & Truptil, S. (2014). Collaborative systems in crisis management: A proposal for a conceptual framework (Vol. 434, pp. 396–405). In: *Proceedings of Pro-VE'14*. Springer, IFIP, Amsterdam, Netherlands.
8. Benaben, F., Lauras, M., Montreuil, B., Faugere, L., Gou, J., Li, J., & Mu, W. (2020). A physics-based theory to navigate across risks and opportunities in the performance space: Application to crisis management. In *Proceedings of 53rd HICSS'20*, Hawaii, USA.
9. Fang, C., Marle, F., Zio, E., Bocquet, J. C. (2012). Network theory-based analysis of risk interactions in large engineering projects. *Reliability Engineering and System Safety, 106*(2), 1–10.
10. Zhang, X., Yang, Y., & Su, J. (2015). Risk identification and evaluation of customer collaboration in product development. *Journal of Industrial Engineering and Management, 8*(3), 928942.
11. Taleb, N. (2007). *The Black Swan—The impact of the highly improbable.* Random House.
12. Ribeiro, J. P., & Barbosa-Povoa, A. (2018). Supply chain resilience: Definitions and quantitative modelling approaches—A literature review. *Computers and Industrial Engineering, 115*, 109–122.
13. Lupton, D. (2013). *Risk* (2nd ed., chap. 2). Routledge, London.
14. Waters, D. (2011). *Supply chain risk management: Vulnerability and resilience in logistics* (2nd ed.). Kogan Page Ltd.
15. Lauras, M., Truptil, S., Charles, A., Ouzrout, Y., & Lamothe, J. (2017). Interoperability and supply chain management. In: *Enterprise Interoperability: INTEROP-PGSO Vision* (pp. 131–151).

Knowledge Representation for Hierarchical and Interconnected Business Contexts

Elena Jelisic, Nenad Ivezic, Boonserm Kulvatunyou, Scott Nieman, Hakju Oh, Nenad Anicic, and Zoran Marjanovic

Abstract Although business context has been introduced as an important concept for message-standards usage and maintenance, its usability depends on the technique used to represent contextual knowledge. This paper investigates a logic-based business-context-modeling technique, which is an alternative technique that can overcome some of the issues identified and discussed in this paper. For other issues, we propose future research directions.

Keywords Business context · Semantics · Enterprise interoperability

1 Introduction

To achieve efficient both integration among enterprise applications and services, and business transactions among trading partners, message standards are needed. Messages define the types of transaction-related information (often referred to as

E. Jelisic (✉) · N. Anicic · Z. Marjanovic
Faculty of Organizational Sciences, University of Belgrade, Belgrade, Serbia
e mail: elena.jelisic@fon.bg.ac.rs

N. Anicic
e-mail: nenad.anicic@fon.bg.ac.rs

Z. Marjanovic
e-mail: zoran.marjanovic@fon.bg.ac.rs

N. Ivezic · B. Kulvatunyou · H. Oh
National Institute of Standards and Technology, Gaithersburg, MD, USA
e-mail: nivezic@nist.gov

B. Kulvatunyou
e-mail: serm@nist.gov

H. Oh
e-mail: hakju.oh@nist.gov

S. Nieman
Land O'Lakes, Shoreview, MN, USA
e-mail: stnieman@landolakes.com

business document specifications) that must be exchanged to achieve both goals. Those standards are the results of the standardization process itself, which can be based on inputs from a large variety of business sectors, business processes, business contexts, and business representatives. Consequently, those approved standards have been only partially successful in achieving their twin goals. There are several reasons. First, the 'out-of-the-box' message standards are agnostic to both specific use cases and implementation languages. Second, the standards typically are not in a digital form that can assure integration and interoperability. Third, the standards become large supersets of data elements contributed by multiple industries. As a result, fourth, detailed refinement of a message standard is necessary to recapture the original business intent and context. Finally, fifth, message standards are traditionally developed in an implementation-specific language, which makes it more costly to deploy a standard to the different platforms and services that exist in every, modern, digital enterprise.

An international team of researchers developed a new software tool, called Score, to address some of those issues [1]. Score is based on ISO 15000 Part 5, an international standard whose goals are (1) to achieve implementation-neutral representation of message standards and (2) to manage separately business context in which the message is used. Score achieves effective management of message-standard profiling. An initial validation of the tool was presented previously in [2].

This paper argues that digitally capturing the business context is critical to improving the usability of existing message standards. Then it points to new research challenges that come with representing that business context in Score. We use the previous Score validation case to identify those challenges. That business case involves a simplified, without consideration of scenario variations, procure-to-pay business process. We used that case study to analyze the current, business-context representation in the Score tool. Based on that analysis, this paper (1) proposes an alternative method for business-context knowledge representation and (2) provides an analysis of its effectiveness. In the newly proposed representation, business context must support both hierarchical and networked knowledge structures.

The rest of the paper is organized as follows. Section 2 introduces background information about important concepts that are used in the paper. Section 3 describes the business-context-definition process in the Score tool and identifies several issues that come with its internal, business-context representation. Section 4 uses the same use case to describe a new, business-context representation. Section 5 discusses the results and proposes future research steps. Section 6 gives concluding thoughts.

2   Background

In terms of messaging standards, a business context (BC) is an information structure whose content characterizes the business transaction for which the standard is devised. That content includes the business process associated with the transaction, each 'entity' participating in the business transaction, their respective 'roles' and their required 'interactions.' An entity is any person, place, or object that is considered relevant to the execution of the business process, which depends on its specific business environment [3].

Score [1] is a novel tool, recently and cooperatively developed by the Open Applications Group Inc. (OAGi) [4] and National Institute of Standards and Technology (NIST) [5]. The tool was used to develop the latest version of the Open Applications Group Integration Specification (OAGIS) standards [6]. The tool has two major benefits. First, it speeds up the development of APIs. Second, it increases the reusability of OAGIS's profiled messages. The tool does so by enriching them with BC, which conveys the intent of the profiled messages.

The problem is that there are various, and sometime incompatible, modeling ways to represent BC. For this paper, we used the analysis in [3], where the author compared six, main, context-modeling techniques: key value, markup scheme, graphical, object-oriented, logic-based and ontology-based. Each technique was analyzed using four criteria: data structure, data-structure components, pros, and cons. The author gave preferences to logic-based and ontology-based models. In this paper, only logic-based models are considered. Future research will consider employing ontology-based models.

Currently, the Score tool represents BC knowledge using graphical, modeling techniques. These techniques, which are based on UML, are good for modeling the structure of BC knowledge. The main advantage of UML models is that they can easily be translated into entity relationship (ER) models—the foundation for database implementations. The main disadvantage of those UML models is the difficulty in using them for reasoning: the main purpose for representing the BC in the first place.

Logic-based models, on the other hand, represent BC knowledge as a set of formal facts that are defined using expressions and a set of rules. The very important advantage of such a representation of knowledge is the support for reasoning processes that can derive new, contextual knowledge by applying rules on those already existing facts. A derived contextual information is represented as a new fact in a formal way [3].

UN/CEFACT's context model (UCM) is an existing application of logic-based modeling technique. In UCM, BC knowledge is represented as a directed, acyclic graph. Conceptually, a BC is an expression created from predicates and logical operators that were applied to specific nodes in those UCM graphs. Each instantiation of such graph represents the possible values in a specific BC context category (e.g., there is an industry, context category and a role, context category). The context values in each graph also have logical, subsumption relationships. Therefore, when using UCM expressions, BC knowledge can be represented more efficiently than by assigning each value individually.

In this paper, we will employ an enhanced UCM model (E-UCM) for the BC knowledge representation that was introduced in [3]. The author gave two important enhancements to the existing UCM model. The first one is the decentralization of the initially centralized UCM graph. This is important, since time needed for graph traversal is directly proportional to the number of nodes and edges. The second one is the introduction of formal definitions of existing operators (intersection (&&),

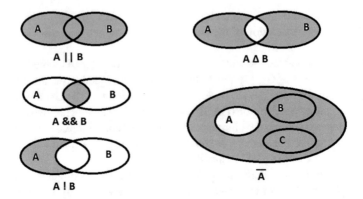

Fig. 1 Operators' representations using Venn diagrams

union (∥) and exclusion (!)) and the introduction of two new operators. The two new operators that improve BC expressiveness are symmetric exclusion (Δ) and complement (Ā). The following figure represents all operators using Venn diagrams (Fig. 1).

3 Definition of Business Context Using the Score Tool: A Use Case

This section introduces a simple use case as a basis for evaluating the current approach for representing BC in the Score tool. First, we describe the profiling process for one, exemplary, business document; then, we identify issues; finally, we give suggestions for improvements.

3.1 Use Case Description

This section describes the simplified, procure-to-pay, business process that we used for analyzing the current, Score tool capabilities. This process describes (1) communication between the customer (aka 'CustomerParty') and the supplier (aka 'SupplierParty') and (2) the respective messages exchanged between these two parties. The messages, which are represented as message flows, include notify shipment, receive delivery, process invoice, and process remittance advice. Figure 2 shows a Business Process Model and Notation (BPMN) collaboration model for the procure-to-pay business process.

Figure 3 shows a BPMN model for the current BC definition process in Score.

As shown in Fig. 3, the process of BC definition can be divided into three main steps—define BC categories, define context schemes for each category, and, put

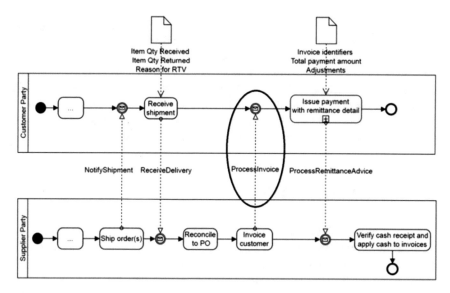

Fig. 2 BPMN model for procure-to-pay business process

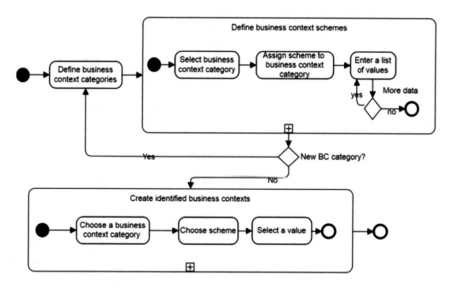

Fig. 3 BPMN model for BC definition process

context scheme values together as required. UN/CEFACT suggests that only eight fixed categories are needed to describe every type of BC. In the Score tool, however, each user can create its own categories that would include BCs based only on what is important to the user. After the BCs are categorized, the respective profile messages can be created in Score and exported in either XML or JSON schema formats. Optionally, the user can choose whether BC definition will appear in the exported schemas. This option significantly increases the potential of using such OAGIS message implementations. Namely, this flexibility makes it possible to have a clear description of each use case, specifying which message profiles are intended to be used. This description is enabled through the BC definition, which improves message profiles' reusability and message-standard maintenance. In this paper, we will focus only on the BC definition, not its impact on message profiles and standards.

3.2   Business-Context Definition

In order to define a BC, you need to know its associated, specific, business processes, and their profile messages. In this paper, we focused on four, business-context categories: *Process classification, business process categories, agricultural industry verticals,* and *organizational structure.* The first category indicates the types of business where the profiled messages can be used. The second category indicates the operational area, within the enterprise, where the profile message can be used. The third one indicates the organizational, operating verticals and business units where the profiled message can be used. Examples include agriculture dairy foods, livestock management, and animal nutrition, among others. The last category indicates the organizational area where the profiled message may be used. For each of the categories, the following context schemes will be used—a food manufacturing enterprise, namely Land O' Lakes, was the operating area for the business process categories; Land O' Lakes business units for the agricultural vertical category; and Land O' Lakes organizational structure for the organizational structure category. APQC PCF cross-reference framework for the process classification category was reviewed but the number of 'levels' caused problems loading into Score. Thus, this category was omitted.

Currently, the Score tool does not support hierarchies; so, twelve context schemes must be created to capture the remaining, three, context categories. For example, the scheme for the business process categories is decomposed into two schemes: namely, the *transportation* scheme and the *shipment request and planning* scheme. *Shipment request and planning* is a node in the *Transportation* scheme; but it also has children nodes. In Score, one scheme can be used for only one category; while one category can take values from multiple schemes. Given these restrictions, the possible BC categories and the assigned schemes are presented in Table 1.

Finally, we have defined one exemplary BC named *shipment request and planning business context.* A part of its definition is presented in Table 2. This is described through *create identified business-contexts* sub-process from Fig. 3. As we can see, it contains three steps. In the first step, the user chooses BC categories that will be used to describe the BC. Afterward, for each category, he chooses one of the schemes and one value at a time to add value to the BC. The list of available values for each

Table 1 Business-context categories and assigned schemas

Business-context category	Business-context schemas
Business process categories	Transportation Shipment request and planning
Agricultural industry verticals	Soil health Crop nutrition manufacturing livestock farming Animal nutrition Water quality/management pest management agronomy services
Organizational structure	Corporate purina dairy

Table 2 Portion of *shipment request and planning business-context* definition

Business-context category	Context scheme	Scheme value
Business process categories	Shipment request and planning	Customer arranged transportation
Business process categories	Shipment request and planning	Supplier arranged transportation
Business process categories	Shipment request and planning	Cross dock
Business process categories	Shipment request and planning	Transfer order
Organizational structure	Dairy	Dairy foods
Organizational structure	Dairy	Vermont creamery
Organizational structure	Dairy	Kozy shack
Organizational structure	Purina	Purina animal nutrition
Organizational structure	Purina	PMI
Organizational structure	Purina	Nutrablend

scheme is already defined in the step *enter a list of values* inside *define business-context schemes* sub-process, as shown in Fig. 3. Through this simple example, the value of BC in enhancing interoperability and reuse is clear. Limitations in Score tool, however, prevent the potential value to be fully realized which are also observed and discussed below.

Limitation 1: Creating BC schemes and their values is a manual task.

While describing the BC for the observed business process and profiled messages, we have created multiple BC schemes and categories. Each scheme is filled manually with individual possible values. In our use case, these schemes did not have a huge number of values. Nevertheless, even in this simple example, it was clear that this can be a time-consuming and error-prone process. In some cases, there are taxonomies that could help define the BC schemes such as APQC PCF. It would be very useful if those taxonomies could be imported in the Score tool automatically, thereby improving its efficiency. Presently, the standards needed to make this possible do not exist.

Limitation 2: No way to express a hierarchy of values within a BC scheme.

Presently, only flat-value structures can be represented in a BC scheme. In other words, there is no possibility to represent parent–child relationships between context values in the Score tool. In some cases, this can be a significant shortcoming, since some BC categories have natural hierarchical organizational structures.

Limitation 3: No way to create associations between values from different schemes.

Associations with defined relationships between values in different schemes are not possible today. For example, from the enterprise perspective, it is important to understand which business units oversee which tasks in the business process. In practice, for example, only dairy foods *support* cross-dock; also, there are identified *synonyms* between different scheme values. Today, creating such associations is not supported in the Score tool.

Limitation 4: BC can be assigned only at the message schema level.

Currently, there is no possibility to assign BCs at the schema-element level. Knowing that there are developed algorithms that in the near future could contribute to message profile creation process, increasing BC granularity may lead to a more productive message profiling.

Limitation 5: BC definition is a manual task. All category values must be assigned individually.

As we saw in Table 2, all values for each BC category must be assigned individually. There is no way to assign these values in a faster and more efficient way. If some category has a huge list of values, this can be a very time-consuming task. In this example, we saw that for *shipment request and planning business-context* category, the organizational structure takes all values from the *dairy* scheme. There is no other way to assign values but to add them separately. In this paper, we propose an alternative BC representation, that could help resolve some of these limitations, leading to a more effective documentation and reuse of message profiles.

4   Business-Context Representation Using E-UCM

This section describes how the same BC can be expressed using the proposed E-UCM technique. The community version of the Neo4j_graph database was used to realize the technique [7]. By analyzing the *shipment request and planning business context* from Sect. 3, we identified three types of nodes: Business-context category, business-context scheme, and scheme values. Consequently, we have created three corresponding labels in Neo4j to make a distinction between these node types. These labels are presented in the upper left corner of Fig. 4. The figure illustrates part of the BC knowledge discussed in Sect. 3, including two BC categories: Organizational structure and business process categories, and the respective three BC schemes. Each scheme has a list of values.

The important observation is the possibility to create child nodes for each *ContextSchemeValue* node. This means that the chosen E-UCM graph representation of BC knowledge supports hierarchies, which is not currently supported in the Score tool. One hierarchy can be identified from Table 2. As shown, *transportation* scheme has six ContextSchemeValue nodes (values). One of these nodes named *shipment request and planning* has a list of its ContextSchemeValue child nodes (*supplier arranged transportation, customer arranged transportation, cross dock* and *transfer order*). Figure 4 includes two associations between nodes from different schemes. The first association, named **supports**, is created between nodes *dairy foods* and *cross dock* (Both nodes have the *ContextSchemeValue* label.). The second association, named **synonym**, is created between nodes *transportation* (label *BusinessContextScheme*) and *logistics* (label *ContextSchemeValue*). In this simple example, we can see that Neo4j_graph database has a natural support for hierarchies and interconnected BC knowledge representations.

By analyzing the BCs presented in Table 2, we have concluded that categories organizational structure and business process categories take all values from the following schemes: *Shipment request and planning, dairy,* and *purina.* Currently, as noted above, each of these values must be added individually and manually to define a BC in Score. We believe that using E-UCM expressions is an easier way to define such a BC. In Table 3, the definition of *shipment request and planning business context* is presented. This table shows that using an E-UCM expression, all values

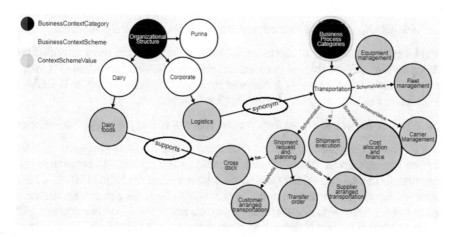

Fig. 4 BC knowledge presentation using Neo4j graph

Table 3 BC expressions for *shipment request and planning business context*

BC expression	Explanation
≤ShipmentRequestAndPlanning	Resolves all successors of the specified BC node including the specified BC node itself
≤Dairy	
≤Purina	
(≤ShipmentRequestAndPlanning) ‖ (≤Dairy) ‖ (≤Purina)	The union of the previous expressions

(nodes) from specified schemes can be resolved by only one simple expression. The combination of these expressions, which is achieved using an existing operator (union, in this example), describes a specific BC (the last row of Table 3). In Neo4j, all expressions are realized using the cipher query language [8].

5   An Analysis of Results

This section presents a detailed analysis of the proposed, alternative technique when it was applied to the BC definition task. We comment on the limitations identified in Sect. 2 and then determine whether the proposed technique can resolve them.

Limitation 1: Creating BC schemes and their values is a manual task.

As stated in the Sect. 3, it would be more productive if the process for defining a BC scheme could be automated since there are existing taxonomies that could be reused. While E-UCM cannot resolve this issue, we propose a future activity to develop a standard for exchanging both the contexts schemes and the corresponding import functionality. When completed, this exchange standard should be added to the Score tool.

Limitation 2: No way to express a hierarchy of values within a BC scheme.

Graphs are a natural way to represent hierarchies. In Fig. 4 we have identified the Transportation scheme that has two hierarchical levels (0. *Transportation* → 1. *Shipment Request and Planning* → 2. *Supplier Arranged Transportation*). In E-UCM, a BC scheme can have as many hierarchical levels as needed.

Limitation 3: No way to create associations between values from different schemes.

In Fig. 4 we have presented two associations (*supports* and *synonym*) that are created between nodes from different schemes. It is important to emphasize that associations can be created between nodes with the same or different labels. In this example, *supports* is an example for an association between nodes with the same label (*ContextSchemeValue*), while the *synonym* is an association between nodes with different labels (*BusinessContextScheme* and *ContextSchemeValue*). Associations' names reveal the nature of nodes' relationships.

Limitation 4: BC can be assigned only at the message schema level.

This cannot be resolved by using E-UCM's BC knowledge representation. Currently, BC can be defined at the component and schema level, but not at the field level. A higher BC granularity can help filter out irrelevant fields. This means that a message profile can be created from bottom-up thereby enabling a pre-profiling process of a component or schema. This process can identify where fields irrelevant to particular BCs would be excluded when querying for a message profile for a BC.

Limitation 5: BC definition is a manual task. All category values must be assigned individually.

This can be resolved only partially with E-UCM. While BC schemes still must be created manually, this task is significantly faster using E-UCM's representation of BC knowledge. This is accomplished using BC expressions that enable more efficient collection of needed values. As presented in Table 3, predicates and operators proposed in E-UCM, and this paper support an easier way of BC definition. We showed that the whole multi-step BC definition in Table 2 was expressed using a single E-UCM expression from the last row in Table 3.

6 Discussion and Next Steps

In this paper, we have identified and analyzed several limitations associated with the BC knowledge representation that is used inside the Score tool. That representation does not support (1) hierarchies and (2) associations between child nodes (specifically between nodes in different context schemes). Furthermore, the values that are used to describe the BC must be assigned individually and manually. In order to resolve these limitations, we have adopted E-UCM, which is a logical technique extended from the prior UCM work for BC knowledge representation. The resulting, graph database has a built-in support for hierarchies and interconnected data coupled with the additional logical operators. Hence, we believe that this approach will prove to be a powerful mechanism for BC definition.

However, E-UCM does not address all of the identified limitations. BC schemes and identified BCs still must be created manually. In addition, there is a problem with BC granularity. As stated, some BC schemes take values from an existing taxonomies; so importing them into a database would make this process more productive. Also, there is a planned integration in the future between the Score tool and the business process cataloging and classification system (BPCCS) [9]. BPCCS enables business process introspection that could provide important information for BC categories (e.g., *business process* BC category).

In our future research, we will consider introducing sensors for BC detection. This means that some categories would not have to have defined schemes (e.g., *geopolitical* BC category). Values for such categories would be detected using virtual or logical sensors [2] (GPS, Web services, etc.). This enhancement would contribute to the automation of processes for defining the BC schemes and the BC.

7 Conclusion

This paper proposes and analyzes a potential technique for knowledge representation that supports hierarchical and interconnected BCs. Through simplified procure-to-pay business process, the paper points at limitations that arise in the current usage of the Score tool—specifically in a business-context, definition process. Currently, the Score tool represents business-context knowledge using graphical modeling technique. New approach proposes logic-based technique, called E-UCM, which is realized through the Neo4j graph database. The paper comments on issues caused by a flat, business-context structure and analyze show the issues get resolved using logic-based knowledge representation. Although E-UCM is a promising avenue, it cannot resolve all the issues identified in this paper. There is an identified need to automate a process of schemes and BC definitions, with an accent on higher BC granularity. Future research will tackle those needs using proposed enhancements. Important conclusion is that BC is a valuable concept, but its usability depends on the technique used to create its representation.

8 Disclaimer

Any mention of commercial products is for information only; it does not imply recommendation or endorsement by NIST.

References

1. Score—Semantic Refinement Tool. http://oagiscore.org/. Accessed 24 October 2019.
2. Jelisic, E., Ivezic, N., Kulvatunyou, B., Anicic, N., & Marjanovic, Z. (2019). A business-context-based approach for message standards use—A validation study. In T. Welzer et al. (Eds.) *New trends in databases and information systems. ADBIS 2019. Communications in computer and information science*, vol 1064. Springer, Cham.
3. Novakovic, D. (n.d.). *Business context aware core components modeling.* Retrieved 28 September 2018 from Publikationsdatenbank der Technischen Universität Wien https://publik.tuwien.ac.at/
4. The Open Applications Group Inc. https://oagi.org
5. The National Institute of Standards and Technology (NIST). https://www.nist.gov
6. OAGIS 10.5 Enterprise Edition documentation. https://oagi.org/
7. Neo4j graph database. https://neo4j.com/. Accessed 10 October 2019

8. The Neo4j Cypher Manual v3.5. https://neo4j.com/. Accessed 10 January2020
9. Ivezic, N., Ljubicic, M., Jankovic, M., Kulvatunyou, B., Nieman, S., & Minakawa, G. (2017). Business process context for message standards. In *CEUR Workshop Proceedings*, 1985, pp. 100–111.

Printed in the United States
by Baker & Taylor Publisher Services